ARCTIC SURVIVAL

MANUAL

Fredonia Books
Amsterdam, The Netherlands

Arctic Survival

by
Ballistic Missile Early Warning System

ISBN: 1-58963-800-X

Copyright © 2002 by Fredonia Books

Reprinted from the original edition

Fredonia Books
Amsterdam, The Netherlands
http://www.fredoniabooks.com

All rights reserved, including the right to reproduce this book, or portions thereof, in any form.

TABLE OF CONTENTS

Section		Page
I	INTRODUCTION	ix
II	THE ARCTIC ENVIRONMENT	1
III	WHAT IT TAKES TO SURVIVE	6

Decision to Stay at the Aircraft or Leave 8
Advantages of Staying With the Aircraft 8
Chances for Rescue 9
Knowledge of Location 9

POSITION FINDING 9

    Latitude by Length of Day 9
    Observations for Latitude 10
    Position Finding by Stars 10
    Longitude from Apparent Noon 11
    Choice of Destination 11
    Condition of Personnel 11
    Radio Fadeout 12
    Magnetic Variation 13
    If You Must Travel 13
    Emergency Signaling 13

TYPES OF SURVIVAL EMERGENCIES 14

    Aircraft Ditching 15
    Survival Emergencies at the Site ... 19
    White Out 20

TABLE OF CONTENTS (cont)

Section		Page
IV	SIGNALING	22
	Radio	23
	Ground Signals	25
	Smoke Flares	29
	Light Signals	30
	Signal Mirror	32
V	TRAVELING IN THE ARCTIC	34
	Decision to Stay or Travel	34
	Exact Knowledge of Present Location and Objective	34
	Knowledge of Orientation Methods	35
	Unusual Physical Stamina	36
	Suitable Clothing	37
	Adequate Food, Fuel, Shelter, or Equipment	37
	Travel	37
	Equipment	38
	Traveling Aids	38
	Visibility	38
	Obstacles	38
	Mountainous Country	40
	Crevasse Rescue	40
	Time to Travel	45
	Timbered Terrain	45
	Rivers and Streams	46
	Barren Land	47

TABLE OF CONTENTS (cont)

Section		Page
VI	FOOD	49
	Animal Food	50
	Large Land Game	51
	Small Land Game	53
	Pole Snare	53
	Sea Ice Game	54
	Hunting Without Firearms	57
	Arctic Birds	57
	Skinning and Butchering	59
	CARE OF MEAT	64
	Protection from Flies	64
	Smoking or Drying Meat	65
	Preserving Cooked Meat	65
	Cooking of Meat	66
	Warm Storage	66
	Sea Food	67
	Winter Fishing	67
	Cleaning and Scaling Fish	70
	Preservation of Fish	70
	Other Edible Marine Life	73
	Cleaning Shell Food	75
	PLANT FOODS	75
	Salmonberry	76
	Mountain Cranberry	77
	Black Crowberry	78
	Bilberry	78

TABLE OF CONTENTS (cont)

Section	Page
Mountain Sorrel	79
Dwarf or Arctic Willow	80
Licorice Root	81
Snake Root	82
Wooly Lousewort	83
Fritillaria	84
Wild Rhubarb	85
Wild Celery or Parsnip	86
FUNGI	87
Gilled Fungi (Mushrooms)	87
Selection of Edible Gilled Fungi	88
Nongilled Fungi	92
Lichens	92
Reindeer Moss	92
Rock Tripe	93
Iceland Moss	98
POISONOUS PLANTS IN THE ARCTIC (TUNDRA) AND SUBARCTIC	98
Baneberry	98
False Hellebore	99
Poison Waterhemlock	101
Vetch, Locoweed Castragalus)	102
Buttercup (Ranunculus)	104
Monkshood, Larkspur	105
Death Camas	106
Water	107
Cooking	108

TABLE OF CONTENTS (cont)

Section	Page
VII CLOTHING AND SHELTER	110
Arctic Clothing	110
Repair Your Clothing	112
Shelters and Shelter Living	112
In the Summer	113
In the Winter	114
Snowhouse or "Igloo"	122
Snow Caves	123
Going In and Out	125
The Snow-Brush is a Lifesaver	125
Cooking Discipline	126
Toilet	127
Beware of Drifting Snow	128
Axes and Knives	128
Firearms	129
Firemaking	130
Useful Hints	136
VIII HEALTH HAZARDS IN THE ARCTIC	138
Freezing	139
Frostbite	139
Flies	141
Carbon Monoxide Poisoning	142
Skin Care	143
Snow Blindness	144
Sunburn	145
Illness	145
Improvised Medical Equipment	145
Severe Chilling	146

TABLE OF CONTENTS (cont)

Section	Page
Immersion Foot (Trench Foot)	146
Tularemia	147
Preventing Infection	148
First Aid	148
Burns	149
Bleeding	149
Cessation of Breathing	152
Artificial Respiration	152
Head Injuries	152
Chest Wounds	154
Shock	154
Eye Injuries	155
Fractures	155
Sprains	156

LIST OF ILLUSTRATIONS

Figure		Page
1.	Position Finding by Stars	10
2.	Ground Air Emergency Code	25
3.	Body Signals	26
4.	Large Dark Against the Snow can be Seen From the Air	28
5.	Panel Signals, Liferaft Sails	28
6.	Standard Aircraft Acknowledgements	30
7.	Signal Mirror Use	33
8.	Formation of a Crevasse	41
9.	Crevasse Rescue	42
10.	Crevasse Rescue	43
11.	Large Arctic Game	52
12.	Pole Snare	54
13.	Sea Ice Game	55
14.	Arctic Birds	58
15.	Where to Make Preliminary Cuts	61
16.	Ice Fishing	69
17.	Preservation of Fish	71
18.	Preservation of Fish	72
19.	Other Edible Sea Life	73
20.	Salmonberry	76
21.	Low Creeping Mountain Cranberry	77
22.	Black Crowberry	78
23.	Mountain Sorrel, A Low Herb	79
24.	Arctic or Dwarf Willow	80
25.	Licorice Root, Height up to 2 Feet	81
26.	Snake Root, up to 10 Inches	82
27.	Wooly Lousewort, up to 8 Inches	83
28.	Fritillaria, Lily-Like Herb, The Roots are Onionlike when Boiled	84

LIST OF ILLUSTRATIONS (cont)

Figure		Page
29.	Wild Rhubarb, A Large Edible Herb, 3-6 Feet High, Found in the Yukon	85
30.	Wild Celery or Parsnip Herb, 2-6 Feet High	86
31.	Poisonous Mushrooms	89
32.	Edible Fungi	90
33.	Boletus	91
34.	Reindeer Moss	93
35.	Rock Tripe	94
36.	Iceland Moss	94
37.	Baneberry (Actaea)	98
38.	False Hellebore (Veratrum)	100
39.	Poison Waterhemlock (Cicuta)	101
40.	Vetch, Locoweed (Astragalus)	103
41.	Buttercup (Ranunculus)	104
42.	Monkshood (Aconitum) and Larkspur (Delphinium)	105
43.	Death Camas	106
44.	Construction of Paratepee	115
45.	Paratepee	116
46.	Easy to Construct Shelters	121
47.	Snow Cave	124
48.	Improvised Stove	130
49.	Waterproof Match Box	131
50.	Cassiope	133
51.	Pressure Points for Bleeding	150
52.	Pressure Points for Bleeding	151
53.	Artificial Respiration	153

INTRODUCTION

Thousands of people every year find themselves confronted with emergency situations which necessitate positive action for human survival. These circumstances are usually the results of unforseen incidents which place human life in a situation where the environment is foreign enough to the individuals involved so as to endanger their health and welfare; climate, food, shelter, etc.

The greatest obstacle that will confront you in the Arctic wilderness or at sea is the fear of the unknown. As you meet and solve such problems you will find that it was not half so bad as you thought it was going to be, and that you are doing pretty well after all.

First, stop and think things over. Size up the situation and plan your course of action. If you are adrift at sea you cannot hurry and there should be an emergency kit and set of instructions in the life raft. On land, however, there will be the temptation to rush off immediately in some direction, any direction, and attempting to do everything at once, thus using up valuable energy and adding to your own confusion. Regardless of the nature of the location in which you find yourself, take time to consider your plight and the best ways to go about improving it.

The vast majority of people who are placed under such circumstances survive and are rescued or find their way back to civilized territory. While the motivation for self-preservation is the strongest motivation we have, it must be remembered that it is not because of the lack of motivation that some unfortunate individuals have perished in a survival situation, but rather through the lack of knowledge or just plain not knowing what to do to overcome the adverse environmental circumstances.

Keeping alive and reasonably healthy in the Arctic environment does not require great physical strength, nor does it involve special skills which are beyond the capabilities of the average man. There are three things which are of the utmost importance; common sense, ingenuity, and discipline and conduct in a survival emergency. This has been proven time and time again by instances of men who have been situated in an unfriendly environment with nothing but their clothes and a few tools, and have survived for months. On the other hand, there have been survival parties which have crash-landed with abundant rations, fuel, clothing, and supplies, and yet have perished in just a few weeks due to lack of leadership, absence of group effort, or lack of specific knowledge to cope with a specific problem.

This booklet, which contains instructions on how to cope with survival problems written by people who have actually lived under survival conditions, tells the main things that a man should know about living in uncivilized regions of the Arctic. Read this book. It may save your life. Keep it in the pocket of your Arctic clothing. With it you may be able to help not only yourself but whoever may be with you.

This book deals with what to do first when an emergency arises, how to make an effective shelter, how to find food, how to keep warm and care for clothing, how to care for sick and injured, and how to signal the rescue party. All of this is taken from the experiences of those who have fought the Arctic and won. Remember, the Arctic can and does support life twelve months of the year, and those who are well prepared for any emergency can win the fight for survival. Do not take the attitude that "it always happens to someone else, and it could never happen to me". Although travel within the Arctic is relatively safe, accidents can happen and weather conditions can change very quickly. The man who thinks that he will never find himself in a survival situation, and consequently, is not prepared, is certainly going to find it more difficult to survive than the one who is well informed and well prepared. It is possible to be caught 50 yards from a building in a "White-Out", and perish the same as if it were a thousand miles from the nearest civilization. Therefore, it behooves you to learn as much as is possible about the techniques of Arctic survival.

In the following pages are numerous hints and suggestions on how to accomplish this.

THE ARCTIC ENVIRONMENT

Most people think of the Arctic as a featureless land of ice and snow, barren and devoid of life. This, of course, is not true. It is true that some parts of it are flat, lifeless and windswept, and covered with ice and snow for most of the year. However, there are also great forests, rugged mountains, lakes, rivers and streams, swamps, and large areas of rolling monotonous tundra. Almost every part of the Arctic supports life in one form or another for all of the year. No part of the Arctic is easy going for man, and in particular, men who are used to a civilization such as ours. Most of Alaska is actually sub-arctic, while a large part of Northern Canada which lies below the Arctic Circle is Arctic in climate. All of Greenland lies above the climatic arctic circle.

Normally, the trip from McGuire AFB to Thule requires 14 hours flying time. It will be seen from the map that most of the first leg is over the mountainous region of eastern Quebec, while practically all of the second leg is over water, except for a small portion of Baffin Island. The route to Site 2, however, will cover almost all types of Arctic terrain — mountains forest, tundra, and sea coast. In view of this, and in view of the fact that a survival emergency can arise anywhere and at any time, all types of Arctic environment will be discussed here.

The tundra is a monotonous, treeless plain. In the winter, it is windswept and barren, and heavy snowfall and extreme temperatures are encountered. In the summer, however, the tundra takes on an entirely different appearance. Forty percent of it is covered by lakes, rivers, and swamps. Shrubs, willows, and numerous other plants grow in abundance. Wildlife is much in evidence, and sufficient food can be obtained from nature. Even in the summer, however, the temperature will go below freezing at night. A survivor who finds himself on the tundra during the summer months will have little trouble finding food or making shelter, but it is a different story in the winter. Game is scarce, and plants must be dug from under the snow. The extreme cold and high winds create an immediate problem. A 20-degree below temperature on a calm day may not bother a man at all, but with a 25 mph wind, exposed parts of the body freeze in 30 seconds. The first concern of a survivor on the tundra in the winter then, would be to make shelter and build a fire for warmth. Fortunately, snow makes an excellent building material, and certain tundra plants will burn, even when wet. A man has one great advantage on the tundra in winter: He, and his camp or the downed aircraft, stand out very well against the snow, making an excellent signal for the rescue party.

Seacoast areas, such as the eastern coast of Canada, are actually part of the tundra, although the terrain is much more rugged. Several advantages will be gained by the man who is forced to survive in a seacoast area. Food is much more plentiful the year 'round, as well as fuel such as driftwood and coal, scraped up in windrows by the sea-ice. There are also habitations on the seacoast, as well as native villages. All natives in the Arctic are friendly, and will help survivors. One problem confronting a survivor on the seacoast would be the fogs, which are prevalent in spring and in autumn, and which would of course, seriously hamper any signalling operations.

The mountains of eastern Canada and Baffin Island are extremely rugged, partly forrested lands, which would present a different set of problems to the survivor. Shelter could be found in a cave, or under a rock ledge, and effective signalling would be comparitively easy. Game, however, is scarce, the only inhabitants of these areas being mountain goats and sheep and birds. These animals are wary and difficult to catch. Other disadvantages would be the difficulty of traveling, even for small distances, and the ever-present danger of snowslides or rockslides. Fuel also may be scarce, unless the survivor is with the aircraft or is fortunate enough to be in a forest area. Herein lies one great advantage of staying with the aircraft — it is an excellent source of fuel, tools and materials.

In the winter, Arctic waters are almost a solid sheet of ice. This ice varies in thickness, but most of it is solid enough to hold a freight train for nine months out of the year. Most Arctic experts agree that a survivor on the sea-ice in the winter stands a better change of survival than anywhere else in the Arctic. The ice is the home of the seal, the major source of food in the Arctic. There are also Arctic fox, birds, and many fish under the ice. The survivor who hunts seals however, must beware, for the seal is also the favorite food of the polar bear, one of the world's most dangerous animals. He will attack man without provocation, and is impossible game without a high-velocity weapon. During the late spring the ice breaks up, and the survivor who finds himself on sea-ice during this time must head for shore and get off the ice. Floes grind and crush against each other with great force, causing pressure ridges as much as 40 feet high. Any man attempting to live on the ice during this time will find himself in danger of drifting to sea, or being crushed between floes. During the summer months, much of the ice is covered with pools of water and slush, and is generally too thin to hold the weight of a man. In the winter, however, the survivor will be able to find ample food, excellent building material, and even fuel (seal blubber) if he is not with his aircraft.

All of the area of northern Canada below the tree line shown on the polar map makes up the arctic forest. Although this area is actually sub-arctic, the temperature in the winter here goes lower than any other place in the Arctic. In the winter there is deep snow, for there is also more snowfall here than anywhere else. There is, however, plenty of game and wood for fuel, and there are rivers and streams which lead to Indian villages. Here again, the natives are friendly and will help survivors. The density of the forest, however, will hamper signalling of the rescue party. In the summer, the forest is hot and there is much rain. Insects will also be a major problem in the summer, as they are in some areas of the tundra. For this reason, survival kits provide headnets and insect repellent to be used in these areas.

There are several important points which it is well to remember about the Arctic, which apply to any area. They are as follows:

> Water is never a problem in the Arctic. This is a great advantage, for a man can go weeks with little or no food, but he cannot last long without water. All water in the Arctic is drinkable. The water in the pools and swamps of the tundra may have a brownish color; none-the-less, it is safe to drink. Ice or snow may always be melted to obtain water, even without a fire by placing the ice or

snow on a dark surface in the sun. Even sea-ice is relatively free of salt, and can be melted and used for drinking water.

All natives in the Arctic are friendly, and will help survivors. Even so, it is best to observe certain precautions when dealing with natives. Respect their customs (they will respect yours as a result), eat sparingly of their food -- they have little enough as it is, and offer payment when you leave. Tobacco is always in demand, and serves as an excellent medium of exchange for services rendered. Avoid close contact with the natives if possible. Americans are common carriers of certain virus diseases which, although harmless to us, are often fatal to Eskimos and Indians. Above all, do not act superior -- you are on the spot, and your life or death may depend upon their help.

Although insects may make a survivor's life miserable if he happens to be on the tundra or in the forest during the summer months, they do not present as serious a problem as in other parts of the world. No Arctic insect carries disease or parasites.

Finally, as was stated previously, no area in the Arctic is easy going for a man who is used to civilization. The survivor will be in a fight for his life the moment he hits the ground, and he may be required to fight harder than ever before.

WHAT IT TAKES TO SURVIVE

When you find yourself in a survival situation, there are several facts which you should remember. One fact is this: the obstacles you have to overcome are not so much natural ones as they are mental ones. Wherever you may be, remember that other people have gone there intentionally, and that some people have chosen to live there. With varying degrees of effort, these people have adjusted to the demands of the terrain, climate, and environment. Your problem is a bit different, however, because you did not expect to be there. So, before you start collecting survival facts and information, you should understand what these psychological obstacles are which you must overcome.

These obstacles all fall under the general heading of that very normal and common emotion called <u>fear</u>. Fear of the unknown -- fear of discomfort -- fear of people -- fear of your own weaknesses. You fear the terrain and the climate because they are new and strange. Because this environment is different, you fear the discomforts which might result. And in many cases, even though these other fears are overcome to some extent, a lack of confidence in their own fortitude and ability has broken people who could otherwise have fared much better.

Though all this is natural, it is not necessary. There are ways of alleviating the needless extra burden that these implanted fears will add.

Your fear of the unknown will be alleviated by learning something of the geography, topography, and climate of the areas you will cross. Learn the methods of getting food and water, and the ways to travel through the terrain.

Learn how to find natural shelter and how to give medical aid, even to yourself, will help to eliminate your fear of discomfort. Most important, you must realize that rest can be more valuable than speed. While making your way across Arctic ice, you will be more successful and comfortable if you make your way with careful planning instead of a blind and exhausting dash.

When the environmental characteristics of the Arctic are recognized, the major requirements for survival become evident. For example, temperatures range from cool to frigid. An inflight emergency may suddenly transfer an airman from a warm aircraft into a barren waste of snow. It is evident that <u>warm clothing</u> is essential to survival.

The same simple reasoning results in emphasizing the need for shelter. With a temperature of -5°F., and a breeze of 8 miles per hour, travel and life in temporary shelter become disagreeable. When it chills to -30°F., and the wind speed doubles, the face will freeze in 1 minute, and travel and life in a temporary shelter become dangerous.

Although the inland areas and the Arctic Ocean are among the calmest localities in the world, strong winds do sometimes sweep along the coasts and across the tundra. Winds of gale velocities may occur where a plateau descends abruptly to the ocean — as along the coasts of Greenland.

It can be seen that food is essential for survival in the Arctic. The colder the weather, the more rapidly heat is dissipated. The source of body heat is the food a person eats. Food is needed to compensate for the accelerated heat loss in cold climates.

DECISION TO STAY AT THE AIRCRAFT OR LEAVE

Your chances of being located and rescued are greatly increased if you can stay with your aircraft and await rescue. Aircraft wreckage is much easier for your rescuers to sight, and it provides many items for your survival and comfort. Have patience; rescue is on the way. Even if you are unable to notify a communications facility of your trouble, skilled rescue personnel will go into immediate action at any time you fail to make a scheduled position report or fail to arrive at your destination. Most rescues have been made when downed crews remained with the aircraft. Leave the aircraft only when:

  a. You are certain of your position and know that you can reach water, shelter, food, and help with available equipment.

  b. After waiting several days, you are convinced that rescue is not coming and you are equipped to travel.

Before making a decision, consider these important points:

Advantages of Staying with the Aircraft

The aircraft is easier to spot from the air than men traveling. Someone may have seen you come down and may be along to investigate.

The aircraft or parts from it will provide you with shelter, signaling aids, and other equipment (use cowling for reflector signals, tubing for shelter framework, gasoline and oil for fires, generator for radio power).

You will avoid the hazards and difficulties of travel.

Chances for Rescue

Your chances are good (1) if you have made radio contact; (2) if you have come down on course or near a traveled air route; (3) if weather and air observation conditions are good.

Knowledge of Location

You must know your location to decide intelligently whether to wait for rescue or to determine a destination and route if you undertake to travel out.

Try to locate your position by studying your maps, landmarks, and flight data, or by taking celestial observations.

POSITION FINDING

Latitude by Length of Day

When you are in any latitude between 60° N and 60° S, you can determine your exact latitude within 30 nautical miles (1/2°), if you know the length of the day within one minute. This is true throughout the year except for about 10 days before and 10 days after the equinoxes — approximately 11-31 March and 13 September - 2 October. During these two periods the day is approximately the same length at all latitudes. To time sunrise and sunset accurately, you must have a level horizon.

Observations for Latitude

Find the length of the day from the instant the top of the sun first appears above the ocean horizon to the instant it disappears below the horizon. This instant is often marked by a green flash. Write down the times of sunrise and sunset. Don't count on remembering them. Note that only the length of day counts in the determination of latitude; your watch may have an unknown error and yet serve to determine this factor. If you have only one water horizon, as on a seacoast, find local noon by the stick and shadow method given below. The length of day will be twice the interval from sunrise to noon or from noon to sunset.

Knowing the length of day, you can find the latitude by using the following nomogram.

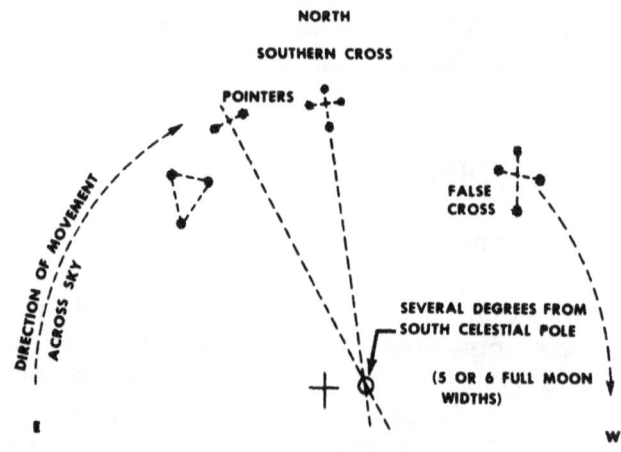

Figure 1. Position Finding by Stars

Longitude from Local Apparent Noon

To find longitude, you must know the correct time. You should know the rate at which your watch gains or loses time. If you know this rate and the time you last set the watch, you can compute the correct time. Correct zone time on your watch to Greenwich time; for example, if your watch is on Eastern Standard Time, add five hours to get Greenwich time.

Choice of Destination

Try to determine the nearest rescue point, the distance to it, the possible difficulties and hazards of travel, and the probable facilities and supplies at the destination.

Condition of Personnel

Consider your physical condition and that of the other men in the party and estimate your ability to endure travel. If there are injured men, try to get help. Send the best fitted men -- two if possible. To travel alone is dangerous.

Before you make a decision, consider all the facts.

If you have decided to stay, then consider these problems:

   a. Your health and body care; the sanitation of your camp.

   b. Your program for rest and shelter.

      c. Your water supplies.

      d. Your food problems.

If you have decided to travel, then these are your problems:

      a. Which direction?

      b. What plan are you following?

      c. What to take along?

Radio Fadeout

Radio fadeouts in the Arctic are caused by solar explosions and sunspot activity. The accepted theory is that the sun emits electrified particles which produce heavy ionization on reaching the earth's atmosphere. This ionized blanket disrupts radio communications everywhere, but particularly in the polar regions. Long term fadeouts may last for several weeks. As these are caused by sunspot activity, they may be forecast.

Short-term fadeouts, caused by solar explosions similar to the detonation of atom bombs, may occur in the Arctic both during daylight and darkness. The atmospheric disturbance is manifest about 18 minutes after a solar explosion. The fadeout condition lasts from 15 minutes to several hours. It cannot be forecast. By blanketing radio reception, fadeouts are of obvious concern to survivors. Radios are unserviceable, and communication leading to rescue may be delayed.

Magnetic Variation

Magnetic variation is noticeable almost everywhere. A compass needle points to true north only from positions due south of the Magnetic Pole, or along the line of no variation on the far side of the North Pole from the Magnetic Pole. The peculiar condition in the Arctic is the enormous variation, up to 180° between the Magnetic and North Poles. A survivor who decides to travel between established points must know that areas of local magnetic attraction exist throughout the Arctic. Navigation based on magnetic compass readings will often be inaccurate.

If You Must Travel

Travel from the crash site should be undertaken only under the following conditions:

If you are sure of your position, know the location of aid, and are physically capable of navigating and surviving until that location is reached, travel may be undertaken.

If you depart the crash site, leave a message giving data and hour of departure and direction of travel. Mark trail at periodic intervals.

If you are in thick forest and jungle areas where an aircraft wreckage may be very difficult to see, proceed to a body of water or open clearing, where your chances of being sighted will be greatly improved.

Emergency Signaling

Familiarize yourself with emergency signaling methods and improvise as many signaling devices as possible. Be prepared to signal at the first sound or sight of a friendly aircraft.

If down in a controlled forest area, start a green brush fire to attract forestry personnel in observation towers. Keep fire under control to prevent spreading.

Make all signal fires downwind from campsite and ground signaling devices to prevent obliterating view from the air.

If sufficient fuel is not available for continuous burning of signal fires, keep fuel dry and ready for instant use. Be sure a combustible material is used to start oil fires.

Immediately upon hearing or sighting a friendly aircraft, transmit a tone signal with your emergency transceiver for automatic homing.

Rescue by Helicopter

If in rough terrain, and landing by helicopters is not possible, proceed to a relatively level surface where the helicopter can hover and pick you up by hoist and sling. Get into the sling in the same way you would put on a coat. Be sure to face the drop cable. If you are injured and unable to get into or stay in sling, a medical technician will be lowered to assist you.

If in a liferaft, use sea anchor and paddles to prevent the liferaft from drifting away because of rotor blade downdraft. Stay in the raft.

TYPES OF SURVIVAL EMERGENCIES

There are two types of survival emergencies which could possibly confront BMEWS site personnel. First of these is an aircraft ditching, which is an extremely remote possibility, but a possibility nonetheless. The second would be it an individual were to be stranded either traveling on foot or by vehicle on the site. Each situation would pose a different set of problems to the survivor. It would be well, then, to discuss the aspects of the two situations to determine just what course of action is required in each instance.

Aircraft Ditching

The term "Aircraft Ditching" usually implies that the aircraft is intentionally pancaked into the sea; however, here we will consider any emergency landing or crash landing as a ditching. All members of MATS crews are thoroughly trained in the techniques of survival, and would be the natural leaders of any survival party. A ditching procedure will be given in the briefings prior to all MATS flights, and will vary slightly according to the type of aircraft. There are certain general points, however, which will apply to any ditching, and it is of value to mention them at this time.

When the pilot decides to ditch, or make an emergency landing, you will be told to fasten the seat belt, or brace yourself against a seat, parachute, or even another passenger. If coming down on water, there are two impacts. The second is the big one — always wait for the second impact before moving. If the plane is coming down on land, wait until it stops moving before doing anything, but get ready to get out. When the plane stops, you start moving and move fast, but keep a clear head. The possibility of a panic is a great danger at this point. The members of the MATS crew will normally direct the exit from the aircraft, leadership being of the utmost importance at such a time. Their directions should be followed without question. You may be told to grab a parachute or a survival kit to take out with you. Injured may need help to get out, but first aid should be postponed until you are clear of the aircraft. If on land, there is immediate danger of fire, and in water the plane will begin to sink as soon as it stops. Therefore, the escape should be made as rapidly as possible, yet order be maintained. As soon as the plane is cleared the fight to survive begins. The course of action that should be taken will depend upon whether the aircraft has come down on land or water.

If the aircraft has ditched in the sea, exposure will be an immediate problem. You will learn during the preflight briefing where the escape hatches are, and where the survival raft is stowed. Remember these locations. The raft should be pushed out first and, of course, inflated AFTER it is outside of the aircraft. Try not to get wet. This may sound impossible, but most large aircraft have an emergency door directly over the wing, inflate the raft and step into it. The danger of exposure will be greatly decreased if clothing and footgear are kept dry. Be careful when entering a survival raft. The bottom is rubberized fabric, and could be torn by the heel of a shoe. Do not jump into the raft. Load any salvaged equipment into it, and step in carefully. Once aboard the raft, the following action should be taken:

Paddle away from the aircraft, avoiding fuel saturated waters, but stay in the vicinity until the plane sinks.
Search for other survivors; account for everyone.
Salvage floating material; check the raft and bail it out.
Rig a windbreak and spray shield. This can be done with a tarpaulin, or a blanket dipped in water and allowed to freeze stiff.
Check condition of survivors; give first aid.
If more than one raft, tie rafts together.
Get emergency radio into operation.

Keep firemaking equipment and compass dry. Place in waterproof container if possible.
Ration water and food. Rig canopy or tarpaulin to catch rainwater or snow.
Start a log and keep an inventory of supplies. Make a calm survey of the situation and plan a course of action.

The last known position of the aircraft will probably be announced when a ditching is imminent, as well as the direction and distance to the nearest land. Survivors should try to make for that land; however, do not try to paddle unless land is in sight. Paddling is exhausting, and if the wind or current (or both) are against you, you will get nowhere. It is best to try to sail, although a raft, having a flat bottom, can be sailed only downwind, or at most, 10° off the wind. To keep the wind from blowing the raft in an undesired direction, a sea-anchor is furnished with the raft, and if used, the raft will drift with the ocean current and be relatively unaffected by the wind. If you are unsure of the direction to the nearest land, or if wind and tide are against you, it may be advisable to stay in the vicinity of the ditching, this being the area in which the search planes will be looking.

Life on a raft is difficult at best. The cramped quarters and lack of activity may create morale problems, as well as the more obvious physical problems of trying to keep warm and dry. Try to keep busy, but do not exert yourself. Tasks should be assigned to every member of the crew. Fish as often as possible, and at different depths. Birds can be shot or caught on a baited hook dragged across the water. If any birds or fish are caught, the entrails should be saved and used for bait. The most important equipment to a raft crew is the emergency radio and the signalling mirror. The radio should be put into use immediately and used constantly. The mirror should be used if an aircraft is seen or heard, or if smoke is seen. More information about the use of these signalling devices will be given later in the course.

The survival party whose aircraft has made an emergency landing on land will generally be in a better position than those who ditched in water. Here it is of extreme importance to clear the aircraft quickly, because of immediate danger of fire. Once clear, set up camp, but remain in the vicinity of the aircraft (makes an excellent signal). When the danger of fire is passed, return and salvage all usable material and fuel from the aircraft. Don't forget to drain the oil in the engine quickly, lest it freeze and remain there. It is useful as fuel and for making signal fires.

The conditions which you find in a survival situation will often depend upon the factors which preceded it. Not all survival emergencies result from a crash landing. An unexpected crackup, a forced landing on a natural airstrip, a bailout, or even isolation from a ground party can result in your being faced with the problem of survival.

An unexpected crash usually indicates the demolition of the aircraft and severe injury or death to many of the crew and passengers. A crash landing or a ditching near shore suggests a hazardous but anticipated landing in which the aircraft is damaged but the crew is fairly intact. A forced landing due to engine trouble or fuel shortage would indicate that the aircraft and its crew were intact but that the aircraft was at least temporarily inoperative. Or, persons or a party involved in a ground operation, such as the installation or operation of a radio or weather station, might easily become separated from the rest of the group and find themselves suddenly faced with a definite survival problem. Bailout from a disabled aircraft can also present such a problem.

Thus, previous events have a definite influence on your ability to cope with the problems of survival. An unexpected crash in which the aircraft is demolished and there are several dead and injured leaves little for the survivors to do but give first aid and make the best of what they have. After a forced or crash landing, shelter, supplies, radio equipment (in some cases), and a number of able-bodied survivors are often available. A bailout or sudden isolation would put a man on his own to a greater degree.

However, unless the contrary is specified, survival procedures outlined in this manual will assume a certain survival situation. They will assume that the aircraft is disabled but that most of the crew is in condition to travel or to forage for food, shelter, and fuel while waiting to be rescued, and that there is a survival kit available. Other situations, such as those involving severe injuries or little equipment, will require modifications in procedures, and these will be specially cited.

Survival Emergencies at the Site

Most of the individual's time at site will be spent indoors; however, during the course of a tour of duty, it will almost certainly be necessary to spend some time outdoors at one time or another. This, plus the fact that vehicular travel from the site to Thule AFB is required, necessitates that prospective site personnel be prepared for a survival emergency which might arise due to vehicle failure or sudden weather.

Site facilities are provided which make vehicle travel as safe as is possible. Over the twelve-mile route from Thule AFB to the BMEWS site, there are four roadside stations placed at intervals. These emergency stations contain a sound-powered telephone connected to the dispatcher's office at the site, a stove, fuel, rations, and other emergency items. In the event of vehicle failure or sudden weather, the individual could almost certainly make his way to one of these shelters until the emergency has passed. In the event that some unforseen emergency arises, the telephone provides an added safety measure. The shelters are also furnished with electrical power from the site.

White-Out

There are two weather conditions which present a danger should a man find himself outdoors — an ordinary snowstorm, and a phenomenon known as a "White-out". A white-out occurs when there is a low ceiling, and light conditions are such that the glare destroys perspective and visibility. Storms and white-outs have two things in common — visibility is zero, and travel is impossible. Usually, warning will be given in advance if a storm is impending, or if a white-out is expected (the latter is harder to foresee). The man who finds himself out-of-doors under either of these conditions should immediately STOP TRAVELING. This weather sometimes lasts for two or three days, so that the man who is caught outside should prepare to wait out the storm. He should establish point of entry, and try to find shelter. Of course, if the man were riding in a vehicle when the weather came up, both of these problems would be solved. If on foot, however, some kind of shelter must be found. If no shelter can be found, sit with your back to the wind with the knees drawn up. Get up and walk a few steps and stamp the feet to retain circulation every so often. In this manner, a man can sit out the storm, and be essentially none the worse, if he is properly clothed. Many an Eskimo has waited out a storm in this way. In any event, do not travel unless you are CERTAIN that there is shelter very close by. If you travel on hunches or vague hopes, you will only become more lost. Conserve your energy and wait out the storm. After a storm, the landscape may be changed considerably. Look for landmarks, such as radio towers, radomes, outstanding topography, etc. Think clearly. Again, do not start to travel unless you are certain of your direction. Listen closely. Sound travels great distances in the cold, dry Arctic air. Vehicles can be heard five miles away, a man stamping his feet two miles away, and ordinary conversation a mile away. Try to backtrack. If you are not certain of which direction to take, it is best to stay where you are and await a rescue party.

With the possibility of a storm or white-out coming up very quickly, it is obvious that certain precautions be taken before going out-of-doors for any reason. First of these is to dress properly. As was stated before, properly clothed, a man has an excellent chance of waiting out any storm which may catch him unawares. Without many layers of clothing to insulate him from the wind and cold, however, there is not much a man can do to help himself. Always play safe — don your Arctic clothing before going out for any reason. The second precaution that should be taken is to inform someone of your leaving, your destination, and when you expect to return.

SIGNALING

The chance of being rescued as soon as possible is enhanced if proper signaling methods are used. Oftentimes, rescue aircraft or rescue parties are sent out, but are unable to sight survivors due to weather conditions, and terrain, even though they may be directly above or within a few miles of the survival party. This is where the lack of knowledge of signaling techniques can be most costly. A rescue plane may be overhead for a minute or so, and you must be prepared to affect a signaling method at but a moment's notice.

There are three types of signaling methods that a survivor may use. He may have all of them, or only one or two of them at his command. None the less, if a survivor, or a survival party, makes a determined effort to attract attention, it is almost sure to be successful. The three types of signals are: radio signals, light signals and ground signals. Of these, radio signals are by far the most effective and require the least effort. However, a radio or electrical power may not be available so that the other types of signals are equally important. Do not depend on radio signals alone. Set up other signals also and keep a 'round the clock signal watch if possible.

The length of time before you are rescued will depend to a considerable extent on the effectiveness of your signals and the speed with which you can have them ready for use. Select the location of each signal carefully and don't set off a holiday fireworks display when you are first spotted. You may need more signals to guide further rescue parties or aircraft. Take care of your signaling equipment.

One man, a group of men or even an aircraft, is not too easy to spot from the air, especially when visibility is limited. Your emergency signaling equipment is designed to make you "bigger" and easier to find.

RADIO

If a pilot knows that he is going to ditch, he will transmit this fact and his position (or last known position) constantly while making his approach. When on the ground, the aircraft radio, if operable, should be put into immediate use. Distress frequencies guarded by all U. S. military units and those of its allies are 3023.5 KC, 121.5 MC and 243 MC. All operating military aircraft have equipment which will cover one or all of these frequencies. On a multichannel equipment one channel is always reserved for a distress frequency and, normally, the equipment must merely be switched to that channel and it is ready to go. Of course, aircraft communications equipments require a battery for power so that the survivor should be certain to protect the battery from freezing or damage at all costs. A good antenna is also required. Make certain that it is not shorted or grounded by snow. To get a good ground for the transmitter go down to the earth - snow and ice make a poor ground for transmission. On sea-ice the ground wire must be submerged in the water below the ice. Transmit often, giving an SOS and your last known position. Be sure to transmit at 15 and 45 minutes after the hour, as these are the times that all units guard distress frequencies exclusively.

All military passenger aircraft are also provided with the AN/CRT-3 (Gibson Girl) emergency transmitter.

This equipment is designed to be used if the equipment on the aircraft is inoperable, or to supplement such equipment. The AN/CRT-3 transmits on the international distress frequency of 500 KC. Power is furnished by means of an internal generator, turned by hand - power. There is an external key, and an automatic internal keying arrangement which will send out SOS signals as the crank is turned. Effective transmission at this frequency requires a long antenna. This antenna is furnished and is raised by either a kite or a balloon. The kite is a silk box kite with an aluminum frame and may be flown in winds as low as 8 knots. If there is no wind, water is added to a canister of chemical which produces hydrogen, inflates the balloon and raises the antenna. Ground is provided by a weight on a ground wire which is designed to be put over the side of a raft or buried. Although cranking the AN/CRT-3 is tiresome, it should be used as much as possible, especially if no other radio equipment is available. Fifteen and 45 minutes after the hour are the best times to transmit on this frequency also.

You may have a URC-4 or URC-11 emergency transceiver in your aircraft or kit. This is housed in a vest and is a UHF-VHF transmitter-receiver with an approximate line-of-sight signal range of 20 miles. With this set you can transmit either voice or cw signals. The receiver is usually adjusted to operate on the same frequency as the transmitter, which is preset to standard distress frequencies of 121.5 megacycles and 243 megacycles.

In all cases, do not raise or fly an antenna during electrical storms.

GROUND SIGNALS

In addition to radio signals, or if no radio equipment is available, ground signals should be used. There are three major points to remember about ground signals and each is a requisite of effective ground signals. The first point to remember is to make them large - at least thirty feet high. The second is to make them geometrical so that there can be no mistake that they are man made. The third is that they must contrast with the landscape so that they can be picked out immediately from the air. Remember you may have little or no idea what the terrain looks like from the air, but if you observe these three rules the signals will be seen if there is anyone around to see them. The type of ground signals which should be used will depend upon the type of terrain. In snow, dark materials are best.

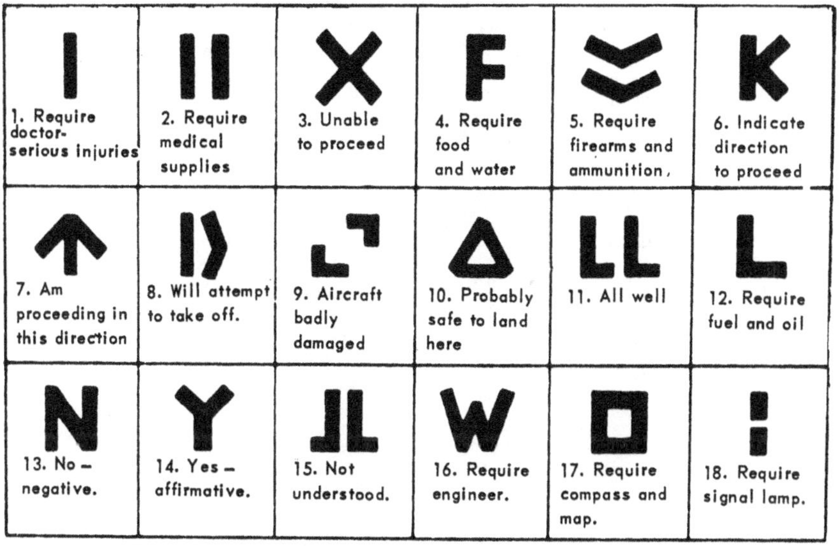

I	II	X	F	≋	K
1. Require doctor-serious injuries	2. Require medical supplies	3. Unable to proceed	4. Require food and water	5. Require firearms and ammunition.	6. Indicate direction to proceed
↑	I)	⌐	△	LL	L
7. Am proceeding in this direction	8. Will attempt to take off.	9. Aircraft badly damaged	10. Probably safe to land here	11. All well	12. Require fuel and oil
N	Y	JL	W	☐	!
13. No – negative.	14. Yes – affirmative.	15. Not understood.	16. Require engineer.	17. Require compass and map.	18. Require signal lamp.

Figure 2. Ground-Air Emergency Code

Branches or stones will serve the purpose well. If none are available, dig a trench. The snow blocks dug out of the trench should be stacked alongside of it - they will cast a shadow and make the trench more noticeable. The aircraft or any shelter you have built from dark material should be kept clean of snow so that they are more easily visible from the air.

Figure 3. Body Signals

Keep snow and frost off aircraft surfaces to make a sharp contrast with the surroundings. A standing spruce tree near timber line burns readily, even when green. Build a "bird nest" of quickly inflammable material in the branches to insure a quick start.

Tramp out signals in snow. Fill them in with boughs, sod, moss or fluorescent dye powder.

In brush country, cut conspicuous patterns in vegetation.

In tundra, dig trenches, turn sod upside down at side of trench to widen signal and cast a shadow.

A parachute tepee stands out in the forest or on the tundra in summer, especially at night with a fire inside.

Make a spruce or conifer torch, selecting a tree with the densest possible foliage. Place dry tinder in the lower branches to light it.

Place or wave the yellow-and-blue cloth signal panel in the open where it can be seen. Spread out parachutes. Make a pattern of orange-colored Mae Wests. Line up cowl panels from engine nacelles upside down on aircraft wings or ground; polish the inside surfaces - as mentioned before, they make good reflectors. Arrange your ground signals in big geometric patterns rather than at random - they will attract more attention that way. If regular panels are not available, use 3 by 12-foot strips from parachute. Rocks, sod, logs or sticks can also be used.

If you are in a wooded area, spread the parachute canopy over the tree tops, if possible.

Figure 4. Large Dark Against the Snow can be Seen From the Air

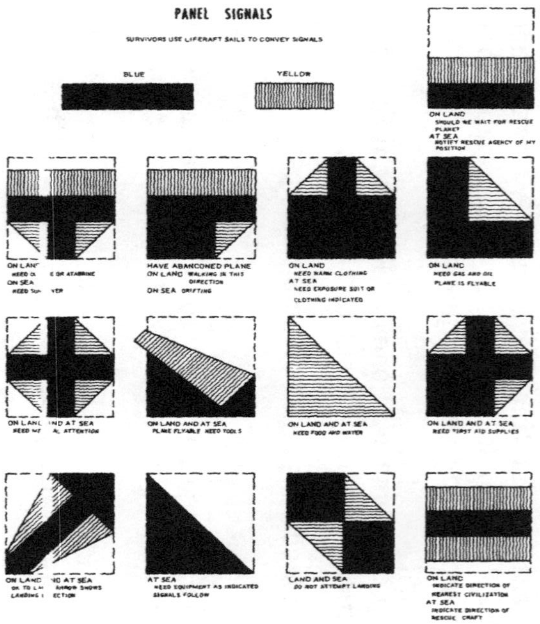

Figure 5. Panel Signals, Liferaft Sails

Note: As sound does not carry well through snow, always keep someone on guard as a spotter. If your entire party is in a snow cave or igloo, you may not hear rescue aircraft. Build the spotter a windbreak but don't roof it.

Use the fluorescent dye available in the life raft or Mae West kit for signaling on water or snow. Use it carefully, for a little goes a long way; use it only downwind for the fine dye will penetrate clothing or food. On rivers, throw it out into the current for a quick spread.

Even if the Gibson Girl radio is inoperative, use the kite alone for a signal. Also, the Gibson Girl balloon can be used as a signal.

Do everything you can to disturb the "natural" look of the ground. If you are down in grass and scrub lands, cut giant markers - a circular path, 8-12 feet in width and 60-75 feet in diameter, is easily seen from the air. A trampled or burned grass pattern will show from the air.

If you can climb a tall tree, hoist a large white or colored improvised flag on a pole lashed to the top.

SMOKE FLARES

When signaling, use smoke by day, bright flame by night. Add engine oil, rags soaked in oil or pieces of rubber (matting or electrical insulation) to make black smoke; add green leaves, moss, or a little water to send up billows of white smoke. Keep plenty of spare fuel on hand.

In general, smoke signals are effective only on comparatively clear and calm days. High winds, rain or snow tend to disperse the smoke and lessen the chance of its being seen. Smoke signals should not be depended on when used in heavily wooded areas and should, if possible, be used in open terrain.

Try to create a smoke which contrasts with its background. Against snow, dark smoke is most effective; against dark turf, white smoke is best. However, to render a sky-silhouetted column of smoke visible to a land rescue party, dark smoke should be used against overcast skies and light smoke against clear skies.

Signaling aids, such as flares and smoke grenades, must be kept dry. Use them only when friendly aircraft are sighted or heard.

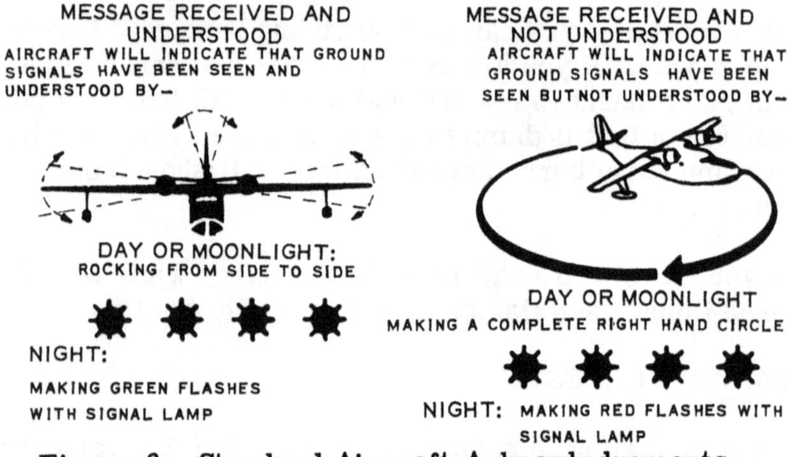

MESSAGE RECEIVED AND UNDERSTOOD
AIRCRAFT WILL INDICATE THAT GROUND SIGNALS HAVE BEEN SEEN AND UNDERSTOOD BY—

DAY OR MOONLIGHT: ROCKING FROM SIDE TO SIDE

NIGHT: MAKING GREEN FLASHES WITH SIGNAL LAMP

MESSAGE RECEIVED AND NOT UNDERSTOOD
AIRCRAFT WILL INDICATE THAT GROUND SIGNALS HAVE BEEN SEEN BUT NOT UNDERSTOOD BY—

DAY OR MOONLIGHT MAKING A COMPLETE RIGHT HAND CIRCLE

NIGHT: MAKING RED FLASHES WITH SIGNAL LAMP

Figure 6. Standard Aircraft Acknowledgements

LIGHT SIGNALS

The third type of signal which the survivor may have at his command is light signals. Light signals are of great value, for if the survivor had no radio and an aircraft was heard during the night the ground signals would be useless. Examples of light signals are a signal fire, flares, flashlight and a signaling mirror.

Several packages of matches in waterproof containers, a lighter and candles are among the most necessary items of your equipment; some should be carried in your pockets and some in the emergency kit. It is much better to take this simple precaution than to be forced later to the difficult and uncertain expedient of making fire by primitive methods. This is particularly applicable to those who may be lost in the Arctic where snow in the winter and rain and mist in the summer make it difficult to obtain suitable dry materials for producing fire.

In order to conserve your supply of matches, light a candle to start the fire. Dry birch bark shredded into strips makes excellent kindling. Use this with wood shavings or dead twigs from standing trees to start the fire. In wet weather, dry fuel may be obtained by cutting into the inner wood of a standing dead tree, or under the surface of a fallen dead tree not resting entirely on the ground. When the fire is burning well, put on green spruce boughs to make a dense smoke.

Outside the forested areas, the best fuel is driftwood. This is usually abundant on northern beaches, though it is scarce in some localities in the central Canadian Arctic. If driftwood is not to be had, dwarf or ground willow will provide an excellent substitute. It grows in protected spots in most parts of the Arctic. In winter it is sometimes found exposed on bare windswept hills or in gullies. The roots will burn as well as the branches. In some varieties of the ground willow, the roots are more extensive than the branches, and when dead and exposed on the surface of the ground make very good fuel.

Another very good fuel is Cassiope, a kind of heather, low-spreading evergreen plant with tiny leaves and white bell-shaped flowers. It grows from 4 to 12 inches high and contains so much resin that it will burn even when green or wet. To get it burning well, however, some dry kindling, such as wood shavings or moss, and a good breeze are required. Dry moss and lichen (plants that grow on the rocks or bark of trees) can also be used as fuel. All these fuels - moss, lichen, heather and willow - can be dug from beneath the snow if necessary.

Signal fires should always be grouped in threes, however separated sufficiently so that they do not appear as one fire from the air. Three fires are an international distress signal. The fires should burn with a flame at night and produce smoke by day. To make black smoke (against snow) use congealed oil or rubber from the tire of the aircraft. To make white smoke (against tundra) burn green branches or any wet vegetation. Flares are furnished with the survival kit. There are bright red flares for night use and smoke flares for day use. They should be saved until an aircraft is actually heard, just as the signal fires should not be lit until you are certain that a rescue plane is in the area. Keeping signal fires going constantly would waste too much fuel, however, they should be ready to light on a moment's notice.

SIGNAL MIRROR

New signal mirrors have been developed for the various survival kits. Follow the instructions on the back of the mirror in your kit. However, there are certain general instructions which hold true for the use of any signal mirror.

Don't continue to flash the mirror in the direction of the rescue aircraft after the receipt of your signal has been acknowledged, unless it appears that the rescuers have lost your exact position. If possible, try to spot the mirror under a wing on the rear of the aircraft. If spotted on the pilot's compartment, it may blind him.

Practice signaling with the mirror in kit. A mirror can be improvised from a ration tin by punching a hole in the center of the lid. An emergency signal mirror can also be made from the aluminum skin of the aircraft A bright polish can be obtained by rubbing dirt over the surface until highly buffed and wiping it off with cloth. Keep the mirror clean. On hazy days, aircraft can see the flash of the mirror before survivors can see the aircraft; so flash the mirror in the direction on the air craft when you hear it, even when you cannot see it.

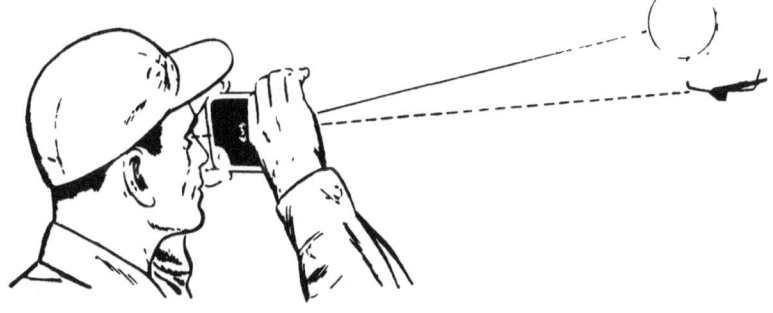

1. REFLECT SUNLIGHT FROM MIRROR ONTO A NEARBY SURFACE (RAFT, HAND, ETC.)
2. SLOWLY BRING MIRROR UP TO EYE-LEVEL AND LOOK THROUGH SIGHTING HOLE. YOU WILL SEE A BRIGHT SPOT OF LIGHT. THIS IS THE AIM INDICATOR.
3. HOLD MIRROR CLOSE THE EYE AND SLOWLY TURN AND MANIPULATE IT SO THAT THE BRIGHT SPOT OF LIGHT IS ON THE TARGET.
4. EVEN THOUGH NO AIRCRAFT OR SHIPS ARE IN SIGHT, CONTINUE SWEEPING THE HORIZON, FOR MIRROR FLASHES MAY BE SEEN FOR MANY MILES, EVEN IN HAZY WEATHER.

Figure 7. Signal Mirror Use

TRAVELING IN THE ARCTIC

Decision: To Stay or To Travel

Travel in the Arctic is extremely difficult and hazardous. The decision to travel should be reached only after careful consideration of the following requirements for successful travel.

    a. Exact knowledge of your present location and of the objective of the journey.

    b. Knowledge of orientation methods.

    c. Unusual amount of physical Stamina.

    d. Suitable clothing.

    e. Adequate food, fuel, and shelter, or the equipment for obtaining them.

These requirements are discussed in the paragraphs which follow.

Exact Knowledge of Present Location and Objective

You must, first of all, have exact knowledge of your location and must know exactly where your objective will be. Establishing a position within a 60-mile square by emergency navigation methods would be helpful to searching aircraft but not exact enough for locating a point of departure for travel on foot. An error of only a few miles in the calculation of travel distance can well be the deciding factor between success and failure.

Knowledge of Orientation Methods

The second requirement is adequate knowledge of orientation methods and means of confirming the course chosen. Natives use natural directional aids to guide them in their journeys. Strangers often wonder at the remarkable sense of direction possessed by comparatively primitive peoples. This wonder is caused by the fact that the stranger fails to recognize the natural aids which the native sees. This fact is as true in the Arctic as it is anywhere else.

Wherever snow falls in the Arctic, the wind whips it into drifts around some object which protrudes above the surface. The largest drifts will therefore be on the lee, or downwind side, of all protruding objects, such as rocks, ice hummocks, clumps of willow, or high banks of creeks. By determining the cardinal points of the compass and the direction of the drifts, you can use the angle at which you cross them as a check point in maintaining a straight course. Drastic changes in terrain features may alter wind direction and thus alter the drifts.

Cloud color is another method of orienting yourself. Clouds over open water will appear dark, while clouds over sea ice appear nearly white. In the early fall, before sea ice is formed, snow-covered land reflects light on low overcast to make it appear nearly white, while clouds over open water will appear dark. Viewed from the land in fall and spring, white sky is known as ice blink. During open winter this reflection is a sure indication of ice.

In the spring, the direction of land or water can often be determined by observing the type and flight of birds. Migrating water fowl, such as ducks and geese, fly toward land in the spring. Seagulls, sea parrots, and auks fly out to the sea ice in the morning and return to land at night, as does the long-tailed jaeger. The jaeger is known to the Eskimos as the doctor bird because of his white breast and neck and black plumage. This bird is easily recognized by his extremely long forked tail feathers and his raucous cry. He has the habit of diving on the Arctic tern in midair to make him disgorge his catch and then seizing it before it has dropped more than a few feet.

Unusual Physical Stamina

The third essential requirement for successful travel is an unusual amount of physical stamina and will to survive. In some cases in which the first and second requirements have been met, overestimation of physical prowess led to fatal results. In relation to this point, it is essential that the traveler realize the need for conserving energy. Survival is synonymous with "take your time". It is not wise to attempt to travel against strong headwinds or to attempt carrying heavy loads for long distances. If you should be caught in a blizzard, make a direction marker. Dig in and remain there until the storm passes. You should be prepared to see a tremendous change in the landscape, following the blizzard. Even with all the desirable equipment, such as snowshoes or skis, and with good weather conditions, Arctic travel is demanding on the physical resources of anyone. Without the necessary equipment and in poor weather, there are a few people who possess sufficient stamina to travel successfully.

Suitable Clothing

Suitable clothing is the fourth requirement for successful travel. If possible, carry clothing in sufficient quantity to give yourself a reasonable chance of remaining dry.

Adequate Food, Fuel, Shelter, or Equipment

The last requirements for successful travel are food, fuel, and shelter, or the equipment which will help you to obtain them from the surrounding environment. Travel consumes body heat at a high rate, and much more food will be necessary while traveling than while remaining relatively inactive in a camp. Therefore, if food supplies are limited and little game is present in the country to be crossed, make certain that travel is the only solution to the problem before starting out.

Note: It is undoubtedly true that the best policy is to remain with the aircraft in the event of an unexpected landing. Since it is impossible to predict the circumstances which may surround a particular emergency in the final analysis, only the survivors can decide whether to travel or remain with the aircraft.

Travel

If decision to travel has been reached after consideration of the requirements discussed in the previous section, make your preparations carefully and equip yourself as best you can. Don't travel in a blizzard or bitterly cold wind--make camp and save your strength until the wind lets up. Don't travel with poor visibility, even if the wind is not blowing.

Equipment

In winter, if you have them available, carry a sleeping bag, parka, mittens, snowshoes or skis, and mukluks. In summer don't forget mosquito netting and repellent, extra clothing (socks especially), and shoepacs. Wear goggles. Keep feet dry, summer and winter.

Traveling Aids

Improvise traveling aids. Make snowshoes from willow branches, aircraft inspection plates, small metal panels, seat bottoms, or metal tubing. Shroud lines and control cables can be used for webbing and harness. Sleds can be made from cabin doors, cowlings, or bomb bay doors. Ropes can be made from parachute lines (each line has about 450 pounds tensile strength).

Visibility

When the sky is overcast and the ground is covered with snow, the lack of contrast makes it difficult to judge the nature of the terrain. In these conditions, men have walked over cliffs before they saw them. Do not travel in these "white out" conditions.

In traveling, remember you are likely to misjudge distances to objects because of the clear Arctic air and lack of familiar scale such as trees and other landmarks. Underestimates of distance are more common than overestimates. Mirage is common in the Arctic.

Obstacles

Obstacles to summer travel are dense vegetation, tough terrain, insects, soft ground, swamps and lakes, and unfordable large rivers. In winter the major obstacles are deep, soft snow, dangerous river ice, "overflows" (stretches of water covered only by a thin layer of ice or snow), severe weather, and a scarcity of native foods. Overflow often lies under deep snow on the surface of the ice and cannot be seen until a person sinks in it.

Because of these hazards, you should seek to travel the barest portions of the river and avoid all obstacles protruding through the ice. Keep to the inside of all curves and away from cut banks where the current is swiftest. As the river progresses to areas where the fall is slackened, the current is slower and the ice is thicker. However, overflow can also occur here, because the ice freezes deeper and blocks the shallower parts of the river bed. Very often after a freezeup the source of the stream or river dries up so rapidly that an air pocket is formed under the initial ice. This is particularly dangerous. Use an ice chisel to test the ice ahead of you.

One danger area still remains on any river anywhere in the Arctic. This is the junction of a creek or other river. Here the resulting whirlpool keeps the water open longer than anywhere else, except swift rapids. In bypassing all junctions of streams, cross well downstream from the mouth of the joining stream. Cross glacier-fed streams early in the morning when the water level is lowest.

When floating down a stream, watch out for "sweepers" — trees that lean nearly horizontally and may brush personnel or equipment off the raft.

Take special care when crossing thin ice. Distribute your weight by lying flat and crawling across. If traveling in a party, rope each man across. If one breaks through, pull him out and get him under shelter at once. Build a fire and dry his wet clothing.

Mountainous Country

In mountainous country it is sometimes best to travel on ridges -- the snow surface is probably firmer and you will have a better view of your route from above. Watch out for snow and ice overhanging steep slopes. Avalanches are a hazard on steep snow-covered slopes, especially on warm days and after heavy snowfalls.

Be especially careful on glaciers. Watch out for crevasses (deep cracks in the ice) that may be covered by snow. Travel in groups of not fewer than 3 men, roped together at intervals of 30 to 40 feet. Probe before every step. Always cross a snow-bridged crevasse at right angles to its course.

Find the strongest part of the bridge by poking with a pole or ice ax. When crossing a bridged crevasse, distribute your weight by crawling or by wearing snowshoes or skis.

Crevasse Rescue

Rescue from a crevasse should be effected promptly.

The first requirement is to firmly anchor the rope by which the man is hanging. The second is to relieve the strain on the hanging man by dropping the other end of the rope or another anchored rope with a loop in the end for him to stand in. This not only facilitates the rescue but also eliminates the serious danger of suffocation from constriction by the rope. He can pass this rope through his chest loop as shown in the illustration. It will serve to prevent his falling backward in

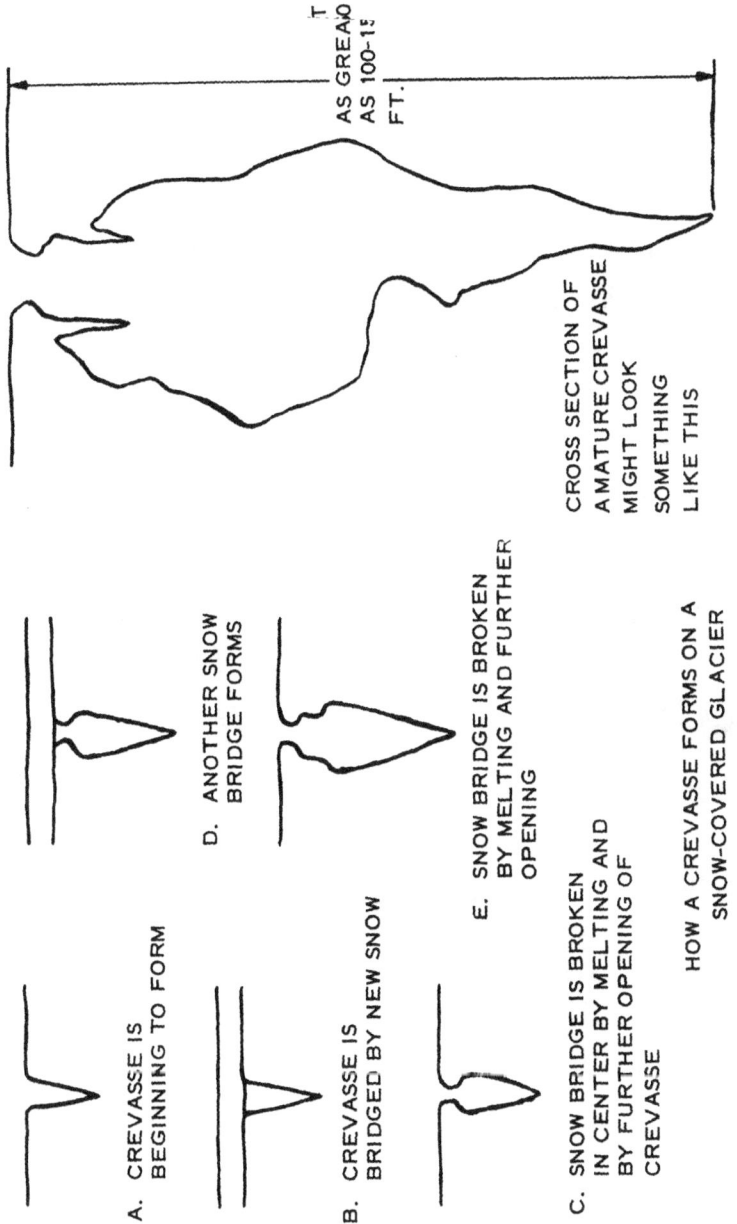

Figure 8. Formation of a Crevasse

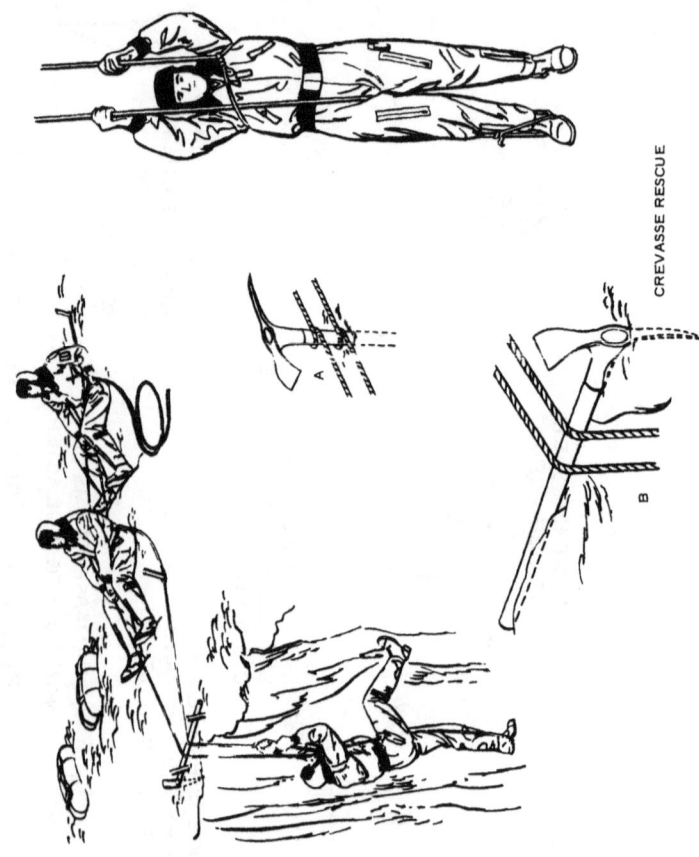

Figure 9. Crevasse Rescue

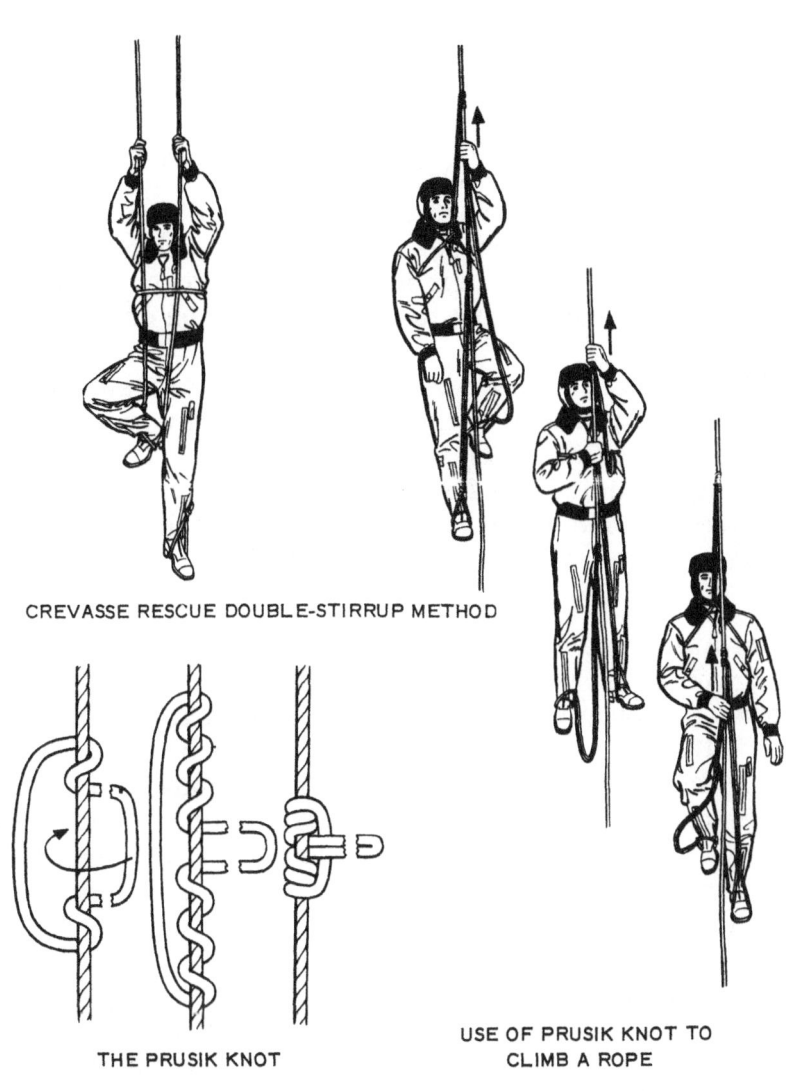

Figure 10. Crevasse Rescue

the ascent. The work will be greatly facilitated if an ice ax or other object can be placed under the ropes at the edge of the crevasse to prevent the rope cutting into the snow, while both ropes have a turn around another ice ax, pole, or solid object for a belay.

The victim now grasps the climbing rope firmly and brings his feet up as high as possible, permitting the man above to take up the stirrup rope. He then stands up and repeats the process, ultimately reaching the lip, where some strong-arm work and a vigorous pull on the rope are necessary to get him over the edge.

If two ends of rope are available, an improvement on this method is to give the hanging man two loops, one for each foot as shown in the illustration. Then by alternately raising one foot and then the other and taking in the ropes at the top, he can get out with less exertion.

Certain precautions should be observed. The exact edge of the crevasse should be ascertained; it may overhang and drop the rescuer into the void if approached too closely. An overhanging lip should be cut away if possible. Otherwise the rope will cut into it, and the man cannot easily be brought over the lip. If the edge is sharp ice, it should be rounded off so that the rope will not be cut. The handle of an ice-imbedded ax can be laid at the edge of the overhang to protect the rope.

The Prusik knot is a valuable means of saving yourself if you fall into a crevasse, and it is useful if you need to bring a man up a rock face. The knot will hold tightly when weight is applied, but it will slide easily when unweighted. The method of tying is shown in the illustration. Take a bight of line and turn it twice around the rope, then pull the loose ends of the line through the loop.

You can also use the Prusik knot for climbing a rope. Use three slings fashioned from lengths of line. Make two stirrups and a chest loop fastened by Prusik knots to the climbing rope as shown in the illustration. There should be about 5 feet between the stirrup and the knot in order to have the knot in front of the chest in an easy handling position. Then, by moving first one stirrup and then the other, it is possible to climb the rope. At the same time, by pushing up the chest loop, you are secured against loss of balance and can take a rest by leaning back at any time.

This is a valuable means of extrication from a crevasse or of rescuing someone from a difficult rock face when strength fails or bad weather intervenes.

Time to Travel

In general, plan to travel during the period from early morning to early afternoon. Make camp early in the afternoon. Then you will have plenty of time to build a shelter and a fire. Dry your clothes, and fix your evening meal, which should be hot and the biggest daily meal. Start again next morning as soon as it gets light.

On the other hand, when you are on glaciers, or or snow-covered terrain in the spring, travel from midnight to noon to avoid runoff streams. Surfaces are better for travel at night, and rest periods are more comfortable during the warmer day. On valley glaciers watch out for falling rocks early in the evening.

Timbered Terrain

The snow lies deep in the timbered areas, and travel is exceedingly difficult without snowshoes or skis.

Progress is generally better when frozen rivers can be followed and the wind has packed the snow hard. However, winter river travel also has its dangers. These are discussed farther on in this section.

North of the timberline major vegetation thins gradually until it finally lies only along creeks and river beds. Here, travel along ridges is often preferable to following rivers unless they are large and fairly clear of snow. In summer, ridge travel is by far the best. The terrain is drier and the ground firmer under foot. Furthermore, if a breeze prevails, it is strongest on the ridges, blowing the mosquitoes down into the valleys and timber.

Rivers and Streams

A glance at any map of the Arctic will reveal that the majority of towns and settlements are located on rivers or on the coast. The reason is obvious. Waterways are the highways of the Arctic. In summer, boats are often the only transportation. In addition, food and fuel are usually available along the inland waterways. Vegetation, brush, and timber line the river banks; fish and bird life are in or on the waters. In short, the natural resources necessary to sustain life are to be found more readily along the rivers, either in winter or in summer.

Rivers which run comparatively straight follow such a course, because fast-flowing current cuts the straightest path through a terrain. The freezing of such rivers is a battle between the temperatures and the current of the river, a battle which never ceases throughout the winter. The current constantly cuts the ice away from below, while the outside temperature tries to maintain or increase the thickness of the covering ice. Snow is on the current's side in this battle, forming an insulation which tends to prevent the outside temperature from reaching the ice. Thus, all snowbanks, especially those on cut banks, are apt to lie on thin ice or no ice at all.

Where gravel bars exist in the bed of the river, overflow often occurs. Such bars freeze solidly and dam the river. The water then seeks an outlet, generally along the bank under a snowdrift or around a log or rock, about which the current is faster.

Barren Land

Winter travel over barren lands is very demanding. Without snowshoes or skis progress can be difficult and slow. Gales, which are impossible to face, may sweep unchecked. There is no natural shelter except that provided by scattered high banks and willow thickets about lakes and along stream beds. Game is very scarce, and fires cannot long be maintained on the fuel generally found in the middle of winter. Most of the rivers follow ancient beds which wind and twist. Survivors who must travel fast cannot afford to follow these old streams which double and quadruple the distance to be covered.

Because of blowing snow, fog, and the lack of landmarks, a compass is a must for barren land travel. Even with a compass, one man has difficulty steering a straight course by himself, and variations in the higher latitudes are extreme. Two can do a little better, but three are best, to navigate when visibility is low. The three progress in single file, the last man carrying the compass. He must always keep the lead man in sight. Using the middle man as a reference point, he controls the course of the lead man by calling out corrections to him. It is strongly recommended that any extended travel over barren land or sea ice be done in groups.

The spring breakup, the summer, and the fall freeze present far greater travel difficulties than does the winter season. Travel must be accomplished on foot without the aid of skis or snowshoes. Equipment must be packed on the back. The masses of soggy vegetation on the tundra cause the traveler to slip and slide. Lake systems must either be crossed or circumnavigated. Use antiexposure suit, if available, in crossing lakes. Be careful in crossing sandbars and mud flats at the mouths and junctions of rivers, and lakes and lagoon outlets. Quicksands and equally dangerous bottomless muck may trap you. Mosquitoes rise in hordes. It is true that there is a much greater possibility of catching fish and shooting birds in summer than there is in winter, but the physical demands of summer travel are more exhausting than those of winter. Rain and fog prevail during breakup and freezeup. Thawing days are followed by freezing nights. The problem of keeping dry, even with rainproof equipment, is almost impossible to solve.

The months of July and August are about the best summer months for cross-country travel. Less rain falls during these months.

If a river flows in your desired direction, float down it rather than attempt to travel cross country. Time spent in constructing a raft will be quickly regained.

FOOD

You will find that the procurement and preparation of foods, especially during the winter months in inland areas of the Arctic, will be your most vital challenge for survival.

Game is most abundant in the summer, but fish can be caught anytime, and plants dug from under the snow can be used for food. There are no poisonous plants north of the tree line and all Arctic animals are safe to eat except for the liver of the polar bear and seal.

One of the first things that should be done by a survival party is to scout the area for signs of game. It is extremely important to start living off the land immediately, and conserve existing rations for emergencies. The presence of game will be indicated by tracks, trails, beds, droppings, or water holes. A 30-calibre carbine and ammunition is provided in the D-1 survival kit, which is effective for game up to the size of a deer. Do NOT, however, attempt to hunt large animals such as polar bear, walrus, moose, or musk ox for you will only wound and anger them, and cause them to attack you as a result. Polar bears are especially dangerous, and should be avoided rather than hunted.

The attitude of the survivor toward procurement of food should be simple: any animal that he can kill is a source of food. Sportsmanship is out. Do not be ashamed at seeking out bait, or shooting birds on the ground rather than on the fly. The survivor's life may depend upon the use of such tactics. Do not waste ammunition on smaller animals if they can be trapped or clubbed.

ANIMAL FOOD

In no part of the Arctic are native animals a reliable source of food. Depending upon the time of the year, and the place, your chances of obtaining animal food will vary. Arctic shores are normally scraped clean of all animals and plants by winter ice. Inland animals are migratory.

Methods Of Hunting

There are basically two methods of hunting with a firearm: stalking and still-hunting. A few points to remember when looking for game are as follows:

    Travel upwind, or crosswind

    Avoid silhouetting yourself on a ridge

    Move slowly and stop often and keep quiet

    Keep cover, even if it means traveling an irregular route

    Look for signs of game

If you are stalking an animal, move only when it is sleeping or feeding. If it looks up in your direction, stand still until it goes back to whatever it was doing. Stay downwind if possible, and try to get close enough for a killing shot. Shoot from the prone position if possible, and rest the weapon on a rock or log for increased accuracy. Shoot for a vital area: head, neck, shoulder or chest. If the animal falls, approach carefully. If you are uncertain, fire a final shot into the head. If the animal runs off, wait a half hour, then follow. Do not follow a wounded animal into a clump of bushes or trees -- they are dangerous at close quarters. If you miss, give the animal a few hours to settle down, then stalk it again. All of this requires time and patience, but is essential if the hunter is to be successful at all.

Still-hunting requires less energy, and is quite often more successful than stalking. The still-hunter should look for a likely feeding place or watering hole and hide downwind. Pick a good hiding spot -- a tree if possible. Move as little as possible while waiting. Try to discover something about the habits of the animals so that you can be there at the most advantageous time. They usually follow a regular routine in feeding and resting.

LARGE LAND GAME

Musk oxen are found in northern Greenland. Caribou (reindeer) may be found in western Greenland. They move close to the sea or high mountains in the summer, while in the winter they feed on the tundra.

Sheep descend to the lower elevations and to valley feeding grounds in the winter.

Wolves usually run in pairs or groups. Fox usually travel alone and are seen when mice and lemming are abundant.

ANIMAL FOOD

Figure 11. Large Arctic Game

SMALL LAND GAME

Tundra animals include rabbits, lemming, mice, ground squirrels, and fox. They may be trapped or shot, winter or summer, anywhere on the tundra. Most prefer some cover and can be found in shallow ravines, or in groves of short willows. Ground squirrels and marmots hibernate in winter. In summer, ground squirrels are abundant along sandy banks of large streams. Marmots live in the mountains, among rocks, usually near the edge of a meadow, or in deep soil - much like woodchucks. To find the burrow in rocky areas, look for a large patch of orange - colored linchen on rocks. This plant grows best on animal or bird dung; and the marmot always seeks relief in the same spot, not far from his well-hidden entrance.

POLE SNARE

Rabbits and ground squirrels can be caught with a spring pole snare. A light pole about 5 feet long (A) is stuck in the ground or a growing sapling can be bent over. Over the runway or the entrance of the burrow is placed a bent-over limb (B), both ends of which are stuck firmly in the ground. A strong cord or light pliable wire is fastened securely to the end of the spring pole, the longer end forming a noose and the shorter end hanging free. The free end and the line leading to the noose are looped into a simple "catch loop". This catch loop holds down the end of the spring pole. After the animal's head goes through the noose, the slightest pull by his shoulders will release the catch loop, thus freeing the spring pole and drawing the noose tight.

If the snare is set to catch rabbits, the bent limb (B) should be about 1 foot 3 inches high at the middle. The bottom of the noose should be about 6 inches off the

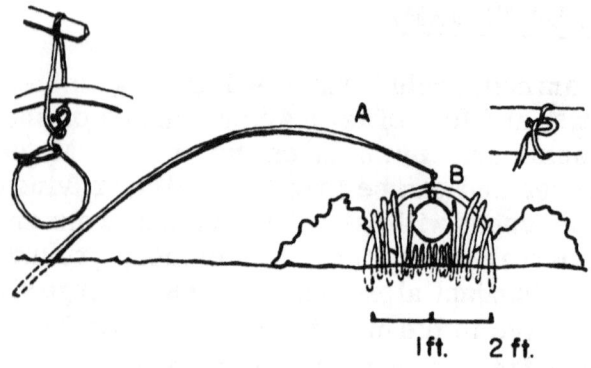

Figure 12. Pole Snare

ground. Brush or limbs should be placed under it and on either side to prevent the animal from going under or around it.

For catching ground squirrels the noose should be placed directly in front of the burrow, and should be slightly smaller than the opening.

If a spring pole is not available, a simple hanging noose can be attached to a fixed limb or other immovable object and suspended over the burrow entrance.

SEA ICE GAME

In the winter and spring, sea mammals -- seals, walrus, polar bears -- are found on the frozen pack ice and on floes in open water. Seals with claws on their flippers can make breathing holes and live relatively close to land. Those that cannot make breathing holes must live on the edge of the pack ice. Most are in groups, but the bearded seal is often found alone.

Figure 13. Sea Ice Game

Seals are hard to approach but every effort should be made to get them for they provide the best meat. In the spring, the seals come up to bask in the sun on the ice beside their breathing holes. They sleep restlessly, raising their heads about every 30 seconds to look around for their enemy, the polar bear. In approaching the seal, crawl forward cautiously while the seal is sleeping, being careful to keep downwind of it. When the seal looks up, lie flat on the ice, imitate its movements by raising your head up and down and wriggling your body slightly. In order to look as much like the seal as possible, approach the seal sideways instead of head-on and keep your arms close to your body.

Since the seal is lying on smooth ice and usually at an incline near the edge of the breathing hole, it must be killed instantly by a shot through the brain, for at the slightest movement of its body it will slide into the water and sink. Therefore, it should be shot through the head at close range, 25 to 50 yards, so that you can dash up and seize it before it reaches the water and sinks.

Seals can also be shot in open water, and in the winter they will usually float. In the winter, the seals come up to breathe at a cluster of holes, each a few inches in diameter. If you can find these holes (they are usually covered with a few inches of snow), you can rig it with a hook to catch a seal. Place a stick across the hole, and tie it to a few feet of line with a hook. Since the holes are only large enough to receive the seal's body, any seal that enters the hole will hook himself and not be able to back out again. The hole will then have to be enlarged to drag out the body. If no hooks are available, you can tie two sticks together in a "T" shape, and put the long stick down into the hole. Stand by the hole, and when the stick moves, drive a spear or harpoon down into the hole. If you can hit the two-inch hole, you can hit the seal. In either of these methods, don't forget to cover the rigged hole with snow, or the light will scare off the seal. When hunting seal, be extremely cautious, because seal is the favorate food of the polar bear, and the hunter may find himself being hunted as well.

Walrus is found on moving ice floes or at leads not far from the shore where they can feed on clams. They should be shot through the neck, just below the head.

HUNTING WITHOUT FIREARMS

If the survivor has no firearm, he is at a disadvantage, but large game can still be killed by other means. A pitfall can be dug, then covered with branches and snow. A fox trap may be rigged in the following manner: Make a cone-shaped structure of snow blocks several feet high with a hole at the top. Place entrails or other bait inside. A fox will climb in the hole to get the bait and be unable to get out. A piece of aluminum from an aircraft a few feet square can be rigged as a deer trap. Cut an "X" in the center of the piece, and place it on a trail, preferably behind a log or some obstacle. The deer will step over the obstacle and put his foot through the metal. The corners will be bent down so that the piece will not come off, and will cause the deer to thrash about to enable the hunter to finish it off. A drag can be made from a noose attached to a large log. The animal will be considerably slowed, having to drag the log through the brush. Other suggestions for hunting without firearms are deadfalls, a slingshot, spear, or at the most elementary level, a club. A man who uses his ingenuity can almost always figure out some way to get food if game is available.

ARCTIC BIRDS

The only birds that remain in the Arctic over the winter are ptarmigan, owls and ravens. Even these are likely to be scarce north of the timber line, especially in the interior. Ravens are thin and of little use as food, but owls and ptarmigan equal any game bird in taste.

Figure 14. Arctic Birds

Ptarmigan are easy to approach, but hard to find, because they are protectively colored in their winter and summer plumages. They may be killed with stones, a slingshot, or a long club. They can also be netted or snared.

In summer, ducks, geese, loons and swans build their nests near ponds on the coastal plains or flats bordering lakes or rivers of the low tundra. A few ducks on a small pond usually indicate that setting birds may be found and flushed from the surrounding shores. Swans and loons normally nest on small grassy islands in the lakes. Geese crowd together near large rivers or lakes. Smaller wading birds customarily fly from pond to pond.

Sea birds may be found on cliffs or small islands off the coast. Their nesting areas can often be located by their flights to and from their feeding grounds. Jaeger gulls are common over the tundra, frequently resting on higher hillocks.

Many birds can be caught with simple noose snares. To catch nesting birds place the noose in the nest to catch the bird's feet. The noose is attached to a stake driven in the ground nearby.

Sea gulls can be caught with a gorge made of a sharp sliver of bone or wood about 3-1/2 inches long with a long line attached at the middle. The bait is a piece of fish or meat completely covering the stick or bone, which, when swallowed, will turn crosswise in the gull's throat.

SKINNING AND BUTCHERING

Under survival conditions, skinning and butchering must be done carefully so that every edible pound of meat can be saved.

The first step in skinning is to turn the animal on its back and with a sharp knife cut through the skin on a straight line, from the end of the tail bone to a point under its neck, A-C on the diagram. In making this cut, pass around the anus and, with great care, press the skin open until you can insert the first two fingers of the left hand between the skin and the thin membrane inclosing the guts. When the fingers can be forced forward, place the blade of the knife between the fingers, blade up, with the knife held firmly in the right hand. As you force forward the fingers of the left hand, palm upward, follow it with the knife blade, cutting the skin but not cutting the membrane. If the animal is a male, cut the skin parallel to, but not touching, the penis. If the tube leading from the bladder is accidentally cut, a messy job and unclean meat will result. If the gall or urine bladders are broken, washing will help clean the tainted meat. Otherwise, however, it is best not to wash meat but to allow it to form a protective glaze.

The following illustration shows preliminary cuts made in skinning and butchering. On reaching the ribs, it will no longer be possible to force the fingers forward, because the skin adheres more strongly to flesh and bone. Furthermore, care is no longer necessary. The cut to point C can be quickly completed by alternately forcing the knife under the skin and lifting it. With the central cut completed, make side cuts consisting of incisions through the skin, running from central cut A-C up the back of each leg to the knee and hock joints. Then make cuts around the front legs just above the knee and around the hind legs above the hocks. Make the final cross cut at point C, and then cut completely around the neck and back of the ears. Now is the time to begin skinning.

On a small or medium-sized animal, one man can skin on each side. The easiest method is to begin at the corners where the cuts meet. When the animal is large, three men can skin at the same time. However, you should remember that when it is getting dark and hands are clumsy because of the cold, a sharp skinning knife

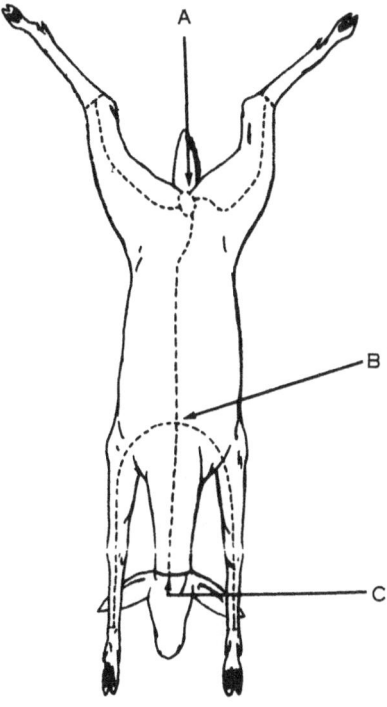

Figure 15. Where to Make Preliminary Cuts

can make a deep wound. So, keep well away from the man next to you. When you have skinned down on the animal's side as far as you can, roll the carcass on its side to continue on the back. Before doing so, spread out the loose skin to prevent the meat from touching the ground and picking up sand and dirt. Follow the same procedure on the opposite side until the skin is free. If you decide before skinning that you do not want the skin, a rough job can be done. However, think well before throwing the skin away. A square of skin, long enough to reach from your head to your knees, will not weigh much when green dried, and is one of the best ground cloths to use under sleeping bag on frozen ground or snow. Snow will not stick to the skin if you lay it hair side up.

Immediately after a kill is the best time to skin and butcher. However, if you kill an animal late in the day, you can make the preliminary cut, A-C, gut the animal, and return early next morning to do the skinning. But be sure to place the carcass so that predators cannot get to it. Then, if the site is visited, the marauder will usually eat only the guts. In opening the membrane inclosing the guts follow the same procedure you followed in cutting the skin, using the fingers of the left hand as a guard for the knife and to separate the intestines from the membrane. You can cut away this thin membrane along the ribs and sides in order to see better. Be careful to avoid cutting the intestines or bladder. The large intestine passes through an aperture in the pelvis. This tube must be separated with a knife from the bone surrounding it. Tie a knot in the bladder tube to prevent the escape of urine. With these steps accomplished, the insides can be easily disengaged from the back and removed from the carcass.

The intestines of a well-conditioned animal will be covered with a lace-like layer of fat, which can be lifted off and placed on nearby bushes to dry for later use. The kidneys are embedded in the back, forward of the pelvis, and are covered with fat. Running forward from the kidneys on each side of the backbone are two long strips of meat called tenderloin or backstrap, which can be cut out. This is eaten after the liver, heart and kidneys. It is usually very tender. Edible meat can also be removed from the head, brisket, ribs, backbone and pelvis. The hams of a large animal are not usually considered good eating. Northern natives frequently feed them to the dogs.

To avoid spoilage, eat the heart, liver and kidneys as soon as possible.

All of the meaty parts of skull such as the brain, tongue, eyes and flesh should also be eaten.

In a severe hunger emergency the intestines, thoroughly cleaned in water, wrapped around a stick and roasted over coals, will be found palatable. The large intestine, cooked in this manner, is considered a delicacy by northern natives.

Remove the bones from the meat. Leg bones laid on a bed of coals will raast quickly and can be easily cracked with light taps of a knife or stone to expose the marrow, which is highly prized as food by all hunters.

When preparing meat under survival conditions, take care not to discard edible fat. This is especially important when, as is often the case in the Arctic, diet must consist mostly entirely of meat. Fat must be eaten in order to provide a complete diet. Many men think that they are unable to eat fat. This is because with a plentiful, civilized diet, meat fat is not a necesity. Under emergency conditions, however, when sugar or vegetable oils are lacking, fat must be eaten. Rabbits lack fat, and the fact that a man will die on a diet consisting of rabbit meat alone indicates the importance of fat in a primitive diet. The same is true of birds such as the ptarmigan.

Birds should be handled in the same manner as other animals. They should be drawn after killing and protected from flies. Birds that carry no fat, such as ptarmigan, crow, and owl, may be skinned. The skins of waterfowl are usually fat and, for this reason, these birds should be plucked and cooked with the skin on. The giblets may also be eaten.

Carrion-eating birds, such as vultures, must be boiled for at least 20 minutes, to kill parasites, before further cooking or eating. Fish-eating birds have a strong, fish-oil flavor. This may be lessened by baking them in mud, or by skinning prior to cooking.

The best meat on a lizard is hind quarters and tail. Eat the legs of a frog. Turtles have flesh on legs, neck, and tail, and tucked away between their shells.

Skin all frogs and snakes. Remove and discard skin, head and viscera.

CARE OF MEAT

Protection from Flies

The greatest danger to meat comes during weather warm enough to allow flies to deposit their eggs, or "blow" the meat. Even while you are skinning an animal, flies can enter bullet holes or any small cavity and lay eggs, which turn into maggots in a few days. The only way of preventing fly blow is to make it impossible for a fly to touch the meat. Do this by wrapping the meat loosely in parachute material or other fabric. Wrap it loosely, so that an airspace of an inch or two is formed between the meat and the sack.

When meat is to be backpacked during the day, it should then be rolled in fabric or clothing and placed inside the pack to be carried. This soft material will act as a nonconductor in keeping the meat cool.

In sparsely settled regions, native dogs will smell meat at incredible distances and raid the meat cache at night. Be careful to guard meat from dogs and other predatory animals.

Smoking or Drying Meat

Cutting meat across the grain in thin strips and either drying it in the wind or smoke will produce "jerky". In warm or damp weather when meat deteriorates rapidly, smoking over a smoldering fire can prevent its spoiling for some time. Take care to keep the meat from getting too hot.

Willow, alders, cottonwood, birch, and dwarf make the best smoking woods. Pitch woods, such as fir and pine, should not be used as they will make the meat unpalatable.

A paratepee would work well for the smoking process. By tying meat to the upper ends good concentration of smoke is obtained. Efficient smoking also can be done by laying fabric over a drying rack and building the fire underneath.

Hang all drying meat high to keep it away from animals. Cover to prevent blowfly infestation. If mold forms on the outside, brush or wash off before eating. In damp weather, smoked or air-dried meat must be redried to prevent molding.

Reptile meat may be dried by placing on hot rocks or hanging in the sun.

Preserving Cooked Meat

To preserve cooked animal food, recook it once each day, especially in warm weather.

COOKING OF MEAT

As mentioned earlier in the chapter, cook all meat thoroughly. All Arctic game, large and small, may harbor trichinosis. Never eat polar bear meat unless it has been thoroughly cooked -- preferably after it has been cut into small pieces and boiled. Polar bear meat is always contaminated with trichinosis. Don't eat the liver of the polar bear or of the bearded seal. It is dangerous to man because of its high concentration of vitamin A. With the exception of marmots and porcupines, no small game is very fat; add available fat when cooking. Remember that rodents may have tularemia -- handle carefully and cook thoroughly.

In selecting the meat to be cooked, remember that the best cuts are found on the head, brisket (breast or lower chest), backbone and pelvis. The hams are likely to be tough and stringy.

WARM STORAGE

In the Arctic in winter you have the problem of warm storage, not cold storage. A game carcass frozen rock-hard at -50° F. is extremely hard to cut unless you have a saw. The alternative is either to cut it into small pieces while it is warm or to insulate it from freezing. A caribou carcass may be kept from freezing for several days by placing the skinned carcass between an envelope of two skins which will seal quickly by freezing along the edges. Then bury this fur bundle in the snow during the night.

Thawed canned rations may be kept from freezing overnight by stowing in the foot of the sleeping bag.

SEAFOOD

Salmon, cod, sculpin, trout, whitefish, herring, flounder and other salt water fish are abundant in the Arctic and along northern Atlantic shores. Some of these can be caught by surf casting from the beach with a long hand line. In spring and summer, salmon enter many of the northern streams to spawn -- often in such numbers that they may be speared from the bank. An improvised gig or spear -- a long pole with two or three sharp wooden barbs lashed to the end -- can be used effectively in shallow water.

Small trout, grayling and other fresh water fish will take any kind of bait -- worm, bug or piece of meat -- and can be caught with the simplest of makeshift tackle, such as a bent pin or small sharp wooden hook attached to a thread or raveling.

The simplest form of hook, one used by primitive peoples in many parts of the world, is the gorge. This is a straight sliver of wood or bone sharpened at both ends and with a line attached at the middle. It is entirely covered with bait and is swallowed lengthwise by the fish. A pull on the line then turns it crosswise in the gullet.

WINTER FISHING

In winter fish can be caught at open leads or through holes cut in the ice*. The hook should be barbless so that the fish can flop off as soon as it is hauled out, for in cold weather it is difficult to remove a barbed hook with bare fingers.

*To keep the hole open, cover it with skins or brush, then heap loose snow over the cover.

Tom-cod and sculpin are the principal salt water fish that can be caught in the winter. Bait is not necessary. A white stone used for a sinker, or a bit of shiny metal or brightly colored cloth tied just above the hook will attract the fish, which can then be caught by jiggling the line up and down.

Poisonous fish are rarer in the Arctic than in the tropics. Some fish, such as sculpins, lay poisonous eggs, but eggs of the salmon, herring or fresh water sturgeon are good eating. In Arctic or subarctic areas, the black mussel may be poisonous at any season. If mussels are the only available food, select only those in deep inlets far from the coast. Remove the dark intestinal gland before eating. Mussel poison is as dangerous as strychnine. Beware of Arctic shark meat, which is poisonous.

River snails or fresh water periwinkles are plentiful in the rivers, streams and lakes of the northern coniferous forest. These snails may have pencil-point or globular shapes, 1 to 3 inches in length. Boil them in water and twist the meat out of the shell with a bent pin or wire.

Shallow lakes freeze to the bottom along the margins, and fish tend to congregate in deeper pools. Estimate the deepest part of the lake or pond before making a hole. Other good locations for ice fishing are at lake outlets or where tributaries flow into a pond or stream. Ice is thinnest over rapids or small falls, or at the edges of deep streams with banks holding drifting snow. Open water is often marked by "smoke", the mist formed by vaporizing water.

In fishing through the ice for lake trout or salmon the hook should be lowered to the bottom, then raised a

Figure 16. Ice Fishing

few inches and kept constantly jiggling. The best bait is a strip cut from the belly of a fish.

In warm weather look for shellfish in muddy waters by feeling the bottom with hands or feet. In deep water, jab a sharp-pointed stick into the slit between the two halves of the shell. When the shells close, pull the shellfish out of the water.

CLEANING AND SCALING FISH

Immediately after you land a fish, bleed it by cutting out the gills and large blood vessels that lie next to the backbone. Scale and wash the fish in clean water.

Some fish, such as members of the trout family, do not need to be scaled. Others such as catfish and sturgeon have no scales but you can skin them.

Some small salt water fish can be eaten with a minimum of cleaning. Their scales are loose and drop off or can be washed off immediately after the fish are caught. The stomach and intestines can be flipped out with the thumb. These fish are oily, highly nutritious, and good -- even raw.

PRESERVATION OF FISH

The method used to preserve fish through several days of warm weather is similar to that used in preserving meat.

When there is no danger of predatory animals disturbing the fish, lay the fish on the available fabric as shown in the diagram. Allow fish to cool all night. Early the next morning, before the air gets warm, turn down the upper edge of the tarp over the top line of fish and turn up the lower edge, B, over the lower line. Then begin

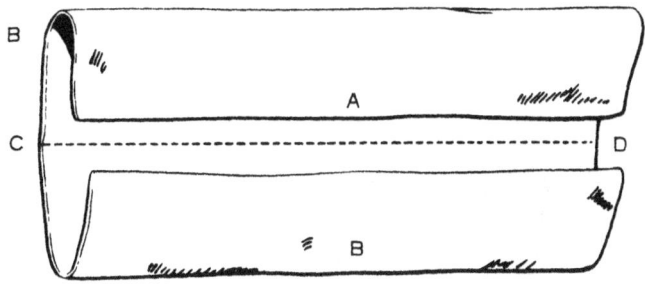
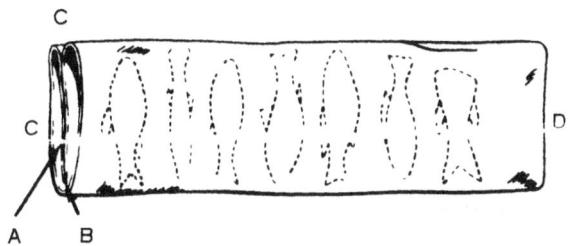
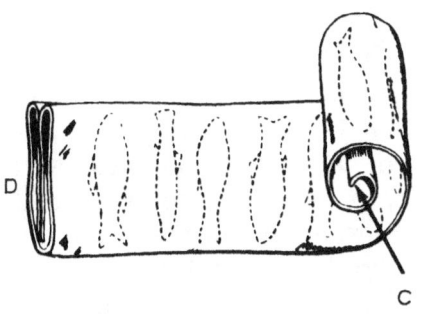

Figure 17. Preservation of Fish

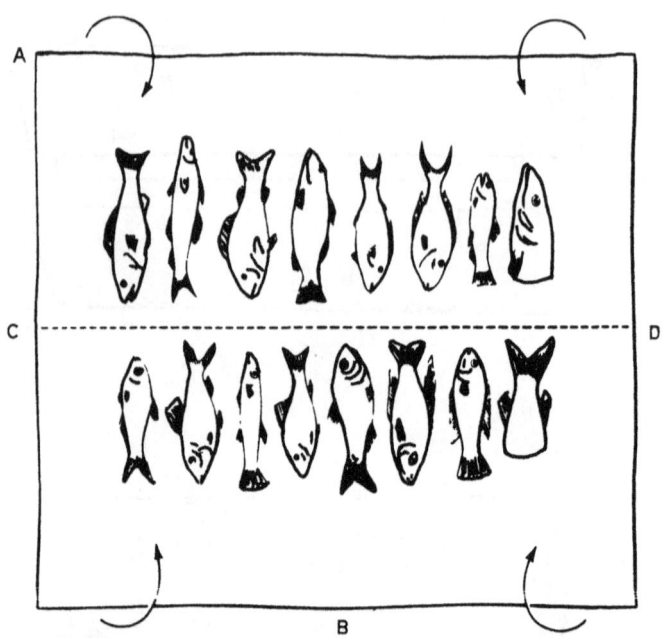

Figure 18. Preservation of Fish

on the edge of the tarp, C, and roll the tarp around the fish until you reach the edge, D. Then, fold the roll thus formed in the center. You will have a rounded roll of protected fish. This roll should be securely, but not tightly tied and wrapped in a sleeping bag, parachute fabric, or clothing, as you would do with meat. This bundle can be placed inside your pack. During rest periods, or whenever the pack is removed, place it in the shade, if possible, to protect it from the direct rays of the sun. If the presence of predatory animals is suspected, suspend the fish from a pole or tree. Cover the package if rain threatens.

Fish may be dried in the same manner as described for smoking meat. To prepare fish for smoking, cut off heads and remove backbone. Then, spread them flat and skewer in that position. Thin willow branches with bark removed make good skewers.

Fish may also be dried in the sun. Hang them from branches or spread on hot rocks. When the meat has dried, splash them with sea water, if available, to salt the outside. Do not keep any sea food unless it is well dried or salted.

OTHER EDIBLE MARINE LIFE

There are many edible varieties of clams, mussels, and other shellfish found in the far North. Most of them live in fairly shallow water. Edible shellfish are relatively abundant on the Arctic and Bering Sea coasts of America. In regions where there is great variation in water level between high and low tides, shellfish can usually be obtained easily at low water, either by digging them with a stick on the tide flats or by collecting them from exposed pools and off-shore reefs. On open

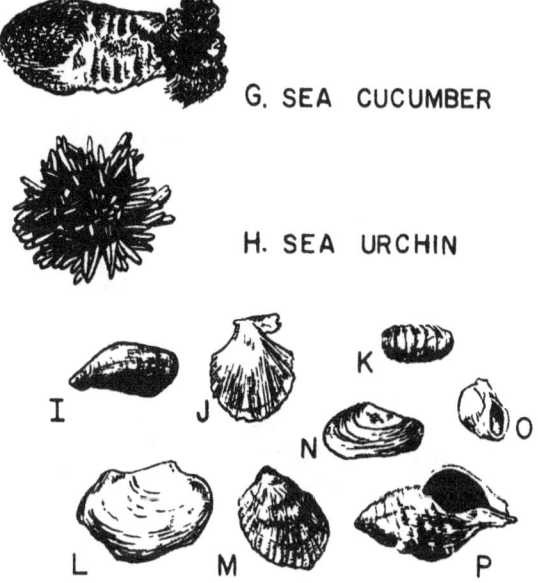

Figure 19. Other Edible Sea Life

sandy beaches with a low tide range, shellfish are often cast up by storm waves. Before eating them, however, make sure they are not spoiled. Generally speaking, the bivalves, such as clams and mussels (I, J, L, M, N) are more palatable than those with spiral shells, though all are edible.

One of the most common edible mollusks of the Far North is the small blackish-purple mussel (I in illustration). In North Pacific waters this mussel is poisonous at certain times of the year and in regions south of the Aleutian Islands they should not be eaten. In the Arctic they may be eaten safely.

Chitons (K in illustration) attach themselves to rocks and have to be pried off. They are oval in shape and have the shell divided into eight separate overlapping plates.

The eggs of the spiny sea urchins (H in illustration) are excellent food. Sea urchins are easily collected among the rocks and in tide pools at low water. The bright yellow eggs or roe are obtained by breaking the shell between two stones. One adult urchin may contain as much as a tablespoon of eggs.

Another sea animal that provides good food is the sea cucumber (G in illustration). Inside the body are five long white muscles that taste much like clam meat.

In early summer it is sometimes possible to scoop up smelt when they come to the edge of the beach to spawn in the surf.

Another valuable source of nourishment can be found in kelp, the long ribbon-like seaweed, as well as the smaller branching variety that grows among the offshore rocks. It is eaten raw. In mid-summer many of the seaweeds are covered with herring eggs.

CLEANING SHELL FOOD

Clams, oysters, mussels, crabs, and lobster left in clean salt water overnight will partially clean themselves and save you some work.

PLANT FOODS

Though plant food is not abundant in the Arctic it is by no means absent. In the summer there are numerous varieties of edible berries, greens, roots and lichens that can be collected along the seacoasts and in the interior if one knows where to look. A few varieties of berries may be found even in winter under the snow.

You will not find any poisonous plants in the Arctic north of the tree line. There are varieties of plants and mushrooms below the tree line that are poisonous, and they will be treated separately at the end of this chapter.

Above the tree line, the only problem is to be able to distinguish the more palatable and nutritious plants from those that have no value as food. Brief descriptions of the principal Arctic food plants are given on the following pages:

The Salmonberry, (Fig. 20), also known as cloudberry, is a low creeping plant, rarely more than three inches high. It is widely distributed in the Arctic, growing

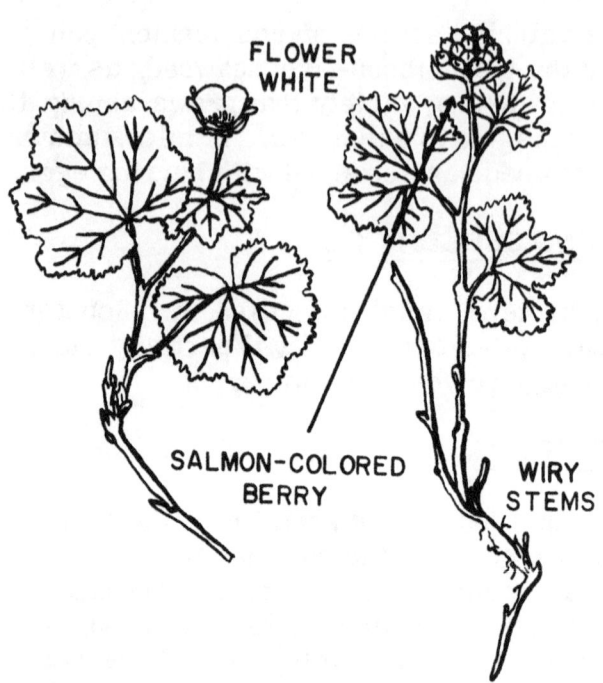

Figure 20. Salmonberry

on mossy, peaty soil. The leaves are large and wide, with five lobes, and the flowers are white, about 3/4 of an inch in diameter. The berries are about the size of raspberries, very juicy with a pleasant taste.

The Mountain Cranberry (Fig. 21) is a low, creeping evergreen shrub from 2 to 8 inches high. It has small, shiny, dark green leaves and clusters of white or pink bell-shaped flowers. The berries are dark red, about 1/4 to 1/2 inch in diameter. They ripen in August and September, but remain on the bush all winter, and can be gathered the next spring when they taste even better than in the fall. The plant is widely distributed in the

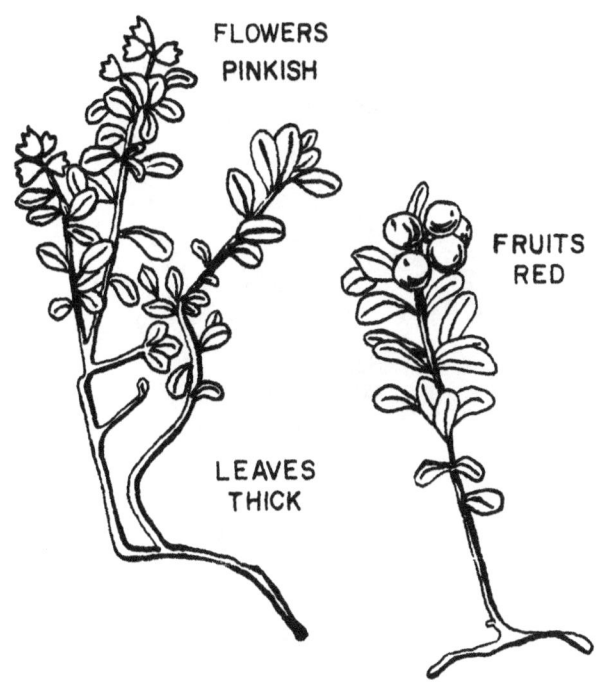

Figure 21. Low Creeping Mountain Cranberry

Arctic, but does not usually bear fruit north of the tree line. It is found in great abundance in open birch and willow thickets.

The Black Crowberry (Fig. 22) is a low evergreen plant with spreading branches and small narrow leaves resembling those of a fir or spruce. The flowers are inconspicuous. The small black shiny berries are sweet and juicy. The crowberry is found in many parts of the Arctic, growing best in sandy or rocky soil, especially along the seacoasts. They can be collected from beneath the snow since they remain on the bush through the winter.

Figure 22. Black Crowberry

Two varieties of Bilberry, closely resembling our blueberry, grow in abundance on the mossy hillsides and tundra. They produce delicious bluish-black berries. Somewhat resembling the bilberries are the Alpine and red bearberry, low plants with small red or black berries. They are rather mealy, and taste better when they are stewed.

Mountain Sorrel (Fig. 23) is a low plant with round or kidney-shaped leaves and stalks of small red or green

Figure 23. Mountain Sorrel, A Low Herb

flowers that grow on shady hillslopes and ravines. The leaves have a pleasant, slightly acid taste and may be eaten raw or boiled.

The willow herb or northern fireweed is an erect plant with dark green, narrow, pointed leaves and large purple flowers that reach a height of 18 inches. It grows on shady or gravelly soil, especially along creeks and river banks. When cooked, the leaves taste something like spinach.

The tender young leaves of the Dwarf Willow (Fig. 24) and also the inner bark of the willow roots may be boiled and eaten.

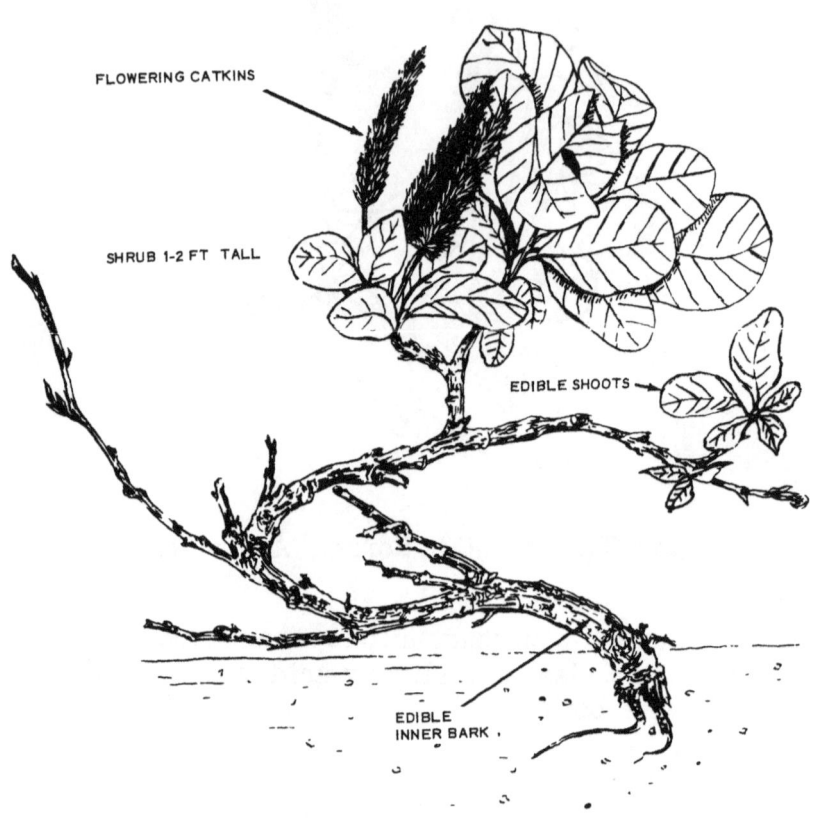

Figure 24. Arctic or Dwarf Willow

Four kinds of edible roots are shown in Figures 25 to 28. Licorice Root (Fig. 25) grows from 1 to 2 feet high and has stalks of pink flowers that develop into seed pods. It has a flexible root, about as thick as a man's finger. When cooked, it tastes like carrots.

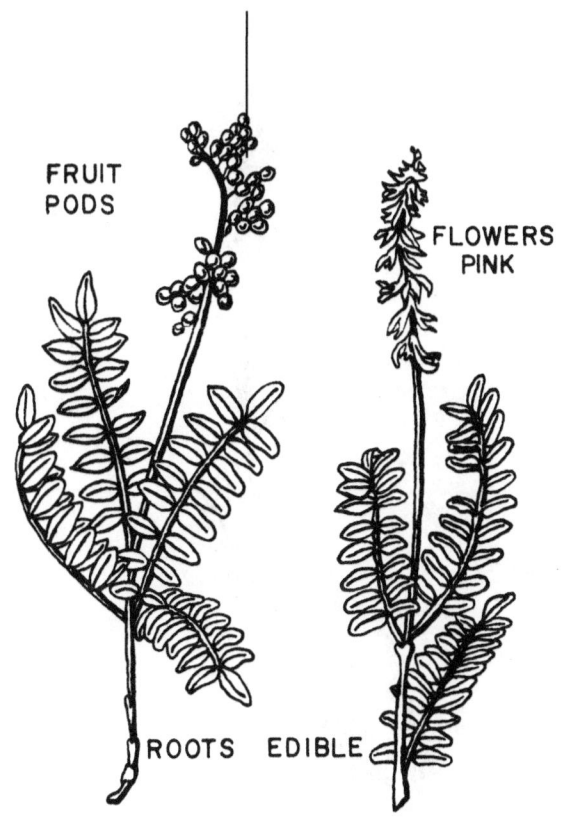

Figure 25. Licorice Root, Height up to 2 Feet

Snakeroot (Fig. 26) from 5 to 10 inches high with large oblong leaves, spikes of white or pink flowers and an edible root about the size of a pecan, grows on dry tundra.

Figure 26. Snake Root, up to 10 Inches

The Woolly Lousewort (Fig. 27) is 5 to 8 inches high with several stems of rose-colored flowers, and has a yellow tap root which tastes like carrots. It is found most in dry tundra country.

Figure 27. Wooly Lousewort, up to 8 Inches

Figure 28. Fritillaria, Lily-Like Herb
The Roots are Onionlike when Boiled

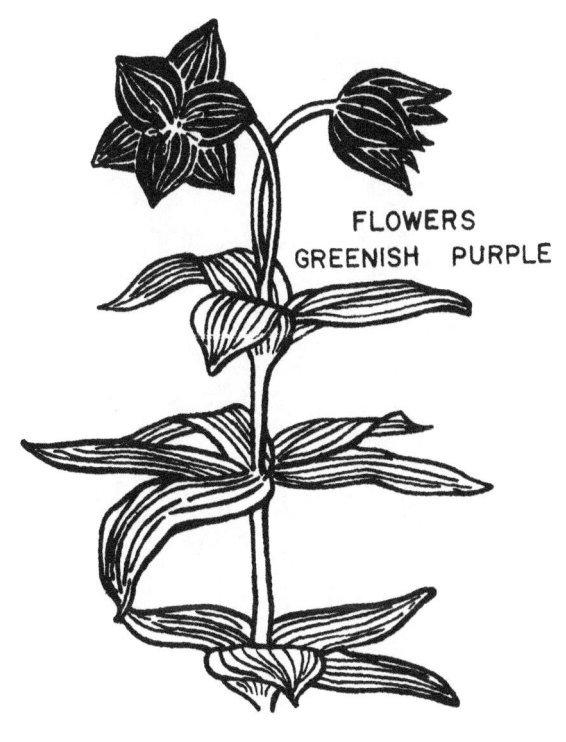

Figure 29. Wild Rhubarb, A Large Edible Herb 3 - 6 Feet High, Found in the Yukon

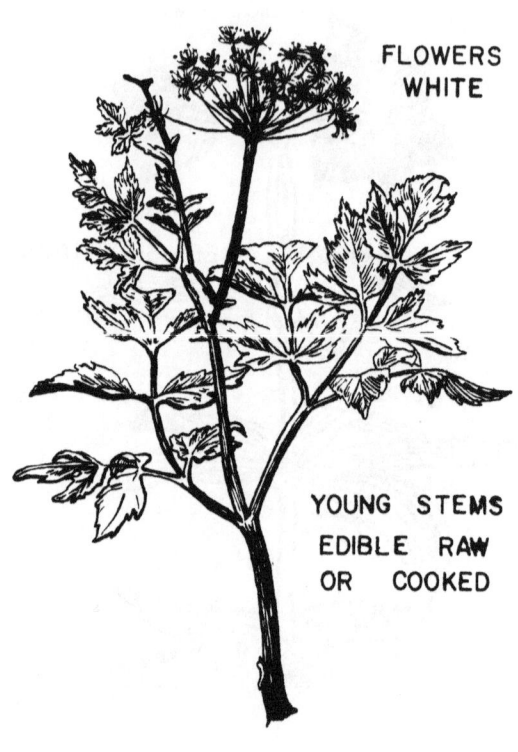

Figure 30. Wild Celery or Parsnip Herb
2 - 6 Feet High

FUNGI

At least 16,000 varieties of edible fungi are known to grow in different parts of the world. Yet, the word "fungi" has an unfortunate connotation to many people - they forget that the mushrooms they eat on their steaks and the moldy part of the blue cheese they spread on their crackers are common forms of fungi. While fungi are not an effective substitute for meat, they do compare favorably with many common leafy vegetables in food content and are often available in many areas where other kinds of edible plants are scarce.

For example, fungi are very common in the wet weather of early spring and late fall in the temperate zones of the world, especially in the pine and spruce-fir forest regions of northern Asia, Europe and North America.

Edible mushrooms can be boiled, baked or used in stews.

GILLED FUNGI (MUSHROOMS)

As mentioned above, mushrooms are the most common of edible fungi, but they have been subjected to an uncommon share of myth and folklore. The term "toadstools" has been used to describe any inedible or poisonous variety of mushroom for so long that people think it is the name of another variety of fungus. Actually, "mushroom" is a more widely accepted name for these plants. The distinguished characteristics ascribed to them by some people are present in edible as well as poisonous types. Odor, peeling of its skin bruises, livid colors - none of these is an acceptable criterion of poisonous mushrooms.

The best method of differentiation is to become familiar with the general characteristics of mushrooms which are edible and of those which are poisonous. The illustrations on the following pages indicate some of the obvious characteristics.

However, a supplementary list of hints to guide you in the selection of edible mushrooms is also indicated below.

Selection of Edible Gilled Fungi

1. Dig the gilled mushroom completely out of the ground before making a decision as to its edibility, for it is especially important to eliminate those mushrooms with a cup, or volva, at the base (See Fig. 32).

2. Avoid all gilled mushrooms in the button state (Fig. 31).

3. Avoid all ground-growing mushrooms with the underside of the cap full of minute, reddish pores (see Boletus, Fig. 31).

4. Avoid all gilled mushrooms with a membrane-like cup or scaly bulbs at the base.

5. Avoid all gilled mushrooms with white or pale milky juice.

6. Avoid all gilled woodland mushrooms with a smooth, flat, reddish top and white gills radiating out from the stem like spokes.

7. Avoid yellow or yellowish-orange mushrooms growing on old stumps. If they have crowded and solid

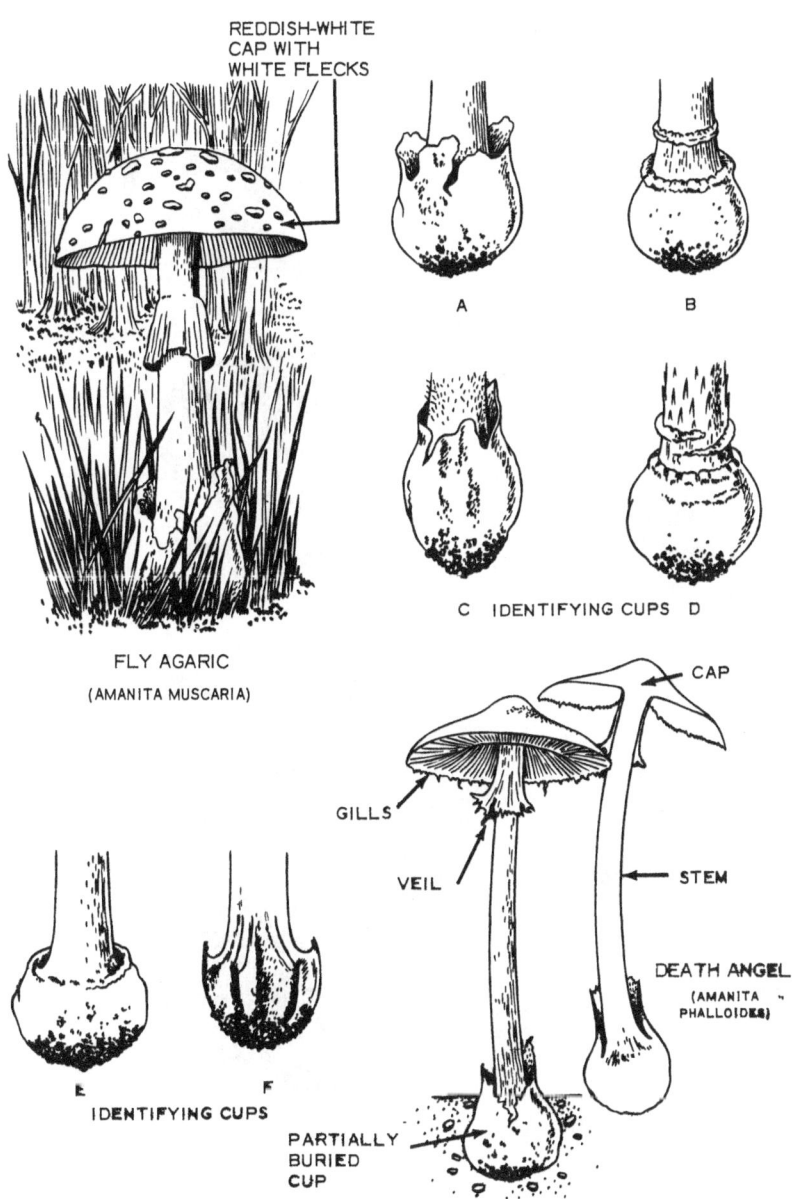

Figure 31. Poisonous Mushrooms

-89-

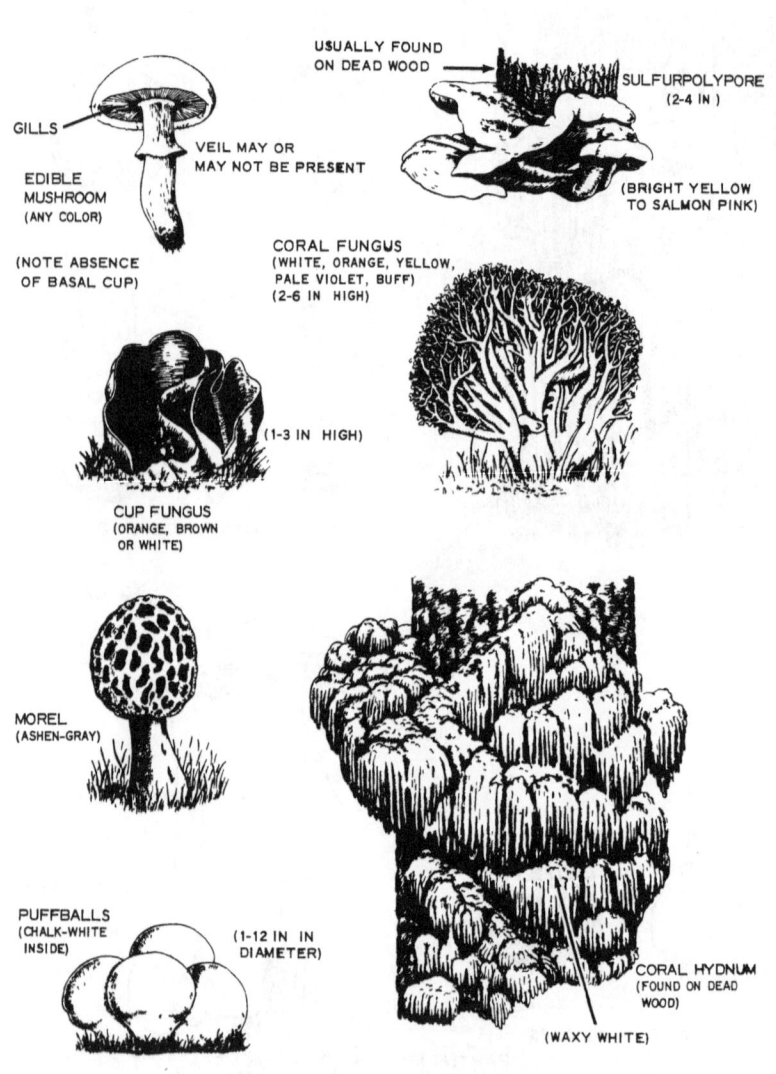

Figure 32. Edible Fungi

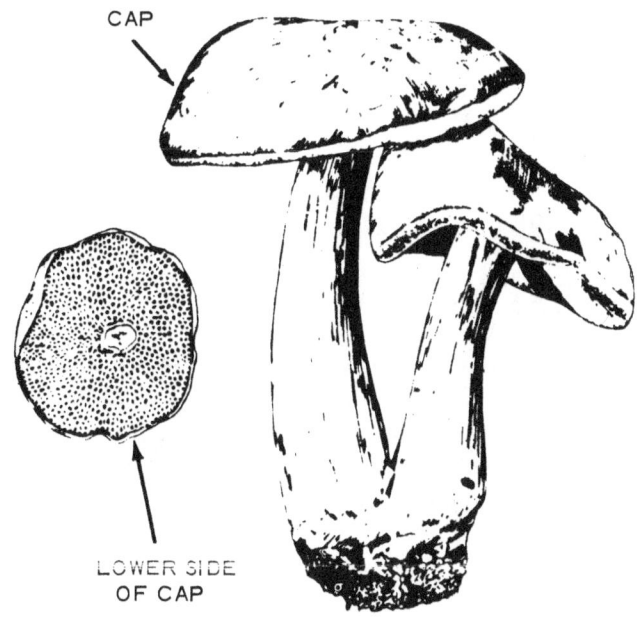

BOLETUS

Figure 33. Boletus

stems, convex overlapping caps, broad gills extending irregularly down the stem, or surfaces that glow phosphorescently in the dark, they are probably poisonous.

8. Avoid <u>any</u> mushrooms which seem be too ripe, water-soaked, spoiled, or maggoty. Any food can be poisonous, or at least inedible when it is decayed or otherwise affected.

If illness occurs after you have eaten some mushrooms, induce vomiting by tickling the back of the throat. Do not take water until after vomiting for it might dilute and spread the poison. However, after vomiting, drink a mixture of lukewarm water and powdered charcoal. Under survival conditions, the only remedy is continued vomiting, "charcoal soup", and rest.

<u>Remember</u> - 98% of all wild mushrooms are edible.

NONGILLED FUNGI

Among the nongilled fungi which grow in abundance throughout the world are the puffballs, morels, coral fungi, coral hydnums and cup fungi. These are illustrated in the drawings in Fig. 32. None of these are poisonous when eaten fresh.

LICHENS

Lichens are abundant and widespread in the far North and can be used as a source of emergency food. However, some of them contain a bitter acid which will cause irritation to the digestive tract.

If lichens are boiled, dried and powdered, this acid is removed and the powder can then be used as flour or made into a thick soup.

<u>REINDEER MOSS</u> (Cladonia rangiferina).

<u>Where</u> found - common in all tundra areas.

<u>Appearance</u> - gray-green and multi-branched.

<u>What</u> to eat - wash the whole plant. Boil or roast it.

GREENISH GRAY.

ON OPEN TUNDRA

Figure 34. Reindeer Moss

ROCK TRIPE (Umbilicaria sp.)

Where found - on certain rocks throughout the northern areas.

Appearance - grayish-black color, leathery and brittle when dry. When it is wet it takes on a dark green color.

What to eat - the whole plant. It will cause diarrhea unless it is dried before it is cooked. Boiling is best.

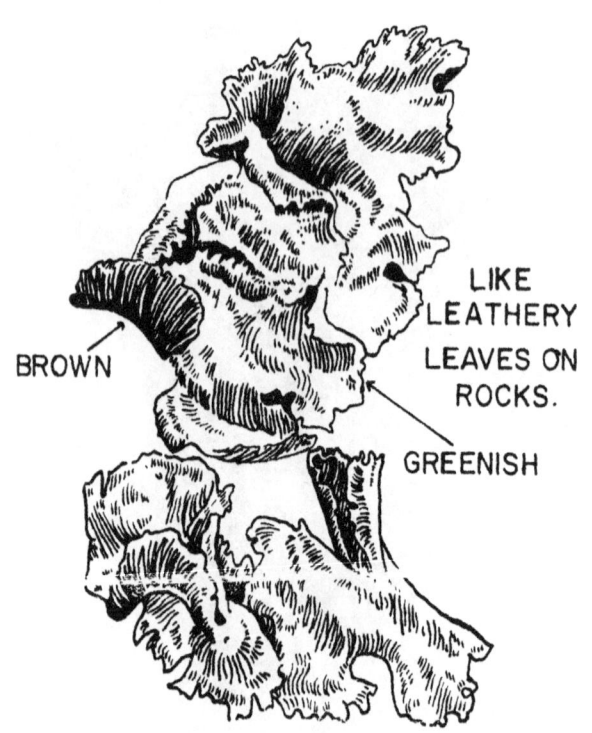

Figure 35. Rock Tripe

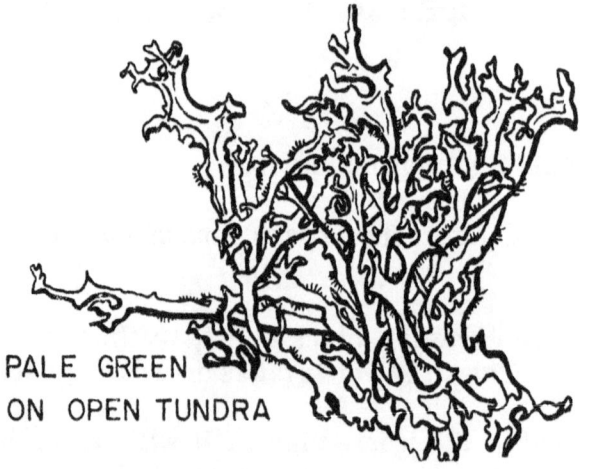

Figure 36. Iceland Moss

PLANT	APPEARANCE	EDIBLE PART	PREPARATION
Artic Willow	The Artic Willow is a shrub which exceeds more than 1 or 2 feet in height. It grows in clumps, which form dense mats on the tundra. Resembles willow tree.	The succulent, tender young shoots of the Arctic Willow can be collected in early spring. The outer bark of the new shoots is stripped off and the inner portion eaten raw. The young underground shoots of any of the various kinds of Arctic willow can be peeled and eaten raw. Young willow leaves are one of the richest sources of vitamin C, containing 7-10 times more than an orange.	Bark...Cooked Shoots...Raw
Wild Blueberries	On the tundra, these wild berries grow on low bushes which are sometimes only a few inches tall. The varieties which grow farther south are produced on taller shrubs which may reach 6 feet in height. The ripe berries are blue, black, or red. These berries are sufficiently common to afford an abundance of fruit during late summer.	WHAT TO EAT. The ripe berries are eaten fresh from the bush or they may be cooked, which makes some kinds more palatable than if eaten fresh. Certain kinds may be dried and eaten like raisins during periods when little else is available.	Raw or Cooked

PLANT	APPEARANCE	EDIBLE PART	PREPARATION
Cloudberry	The ripe cloudberry grows on an erect plant which seldom grows over a foot in height in the southern limit of its distribution and only a few inches tall on the Arctic tundra. The plants have a few undivided, rounded, scalloped leaves. The fruit is borne at the top of the plant, first pink, then amber, at last yellow and very juicy and soft.	The soft ripe fruit has a flavor strongly suggestive of poorly flavored baked apples. A taste must be developed for the cloudberry, but once acquired, large quantities can be consumed in season, usually late summer. The cloudberry is only one of several kinds of wild berries related to the blackberry, raspberry, and dewberries, which the survivor can expect to find in northern regions.	Raw
Lichens	Gray or white stems	All	Boiled or soaked, crushed, and boiled again
Licorice Plant	1-2 ft. high, purple flower, flat seed-pods	Root	Cooked (Aug.-Nov.)

PLANT	APPEARANCE	EDIBLE PART	PREPARATION
Bistort	5-10 in. high, pink or white flowers	Flower, root, leaf buds	Soak and cook roots
Wooly Lousewort	5-8 in. high, pink or purple flower, yellow root	Root	Cooked or raw (late summer)
Red Bearberry	Bush, single red berries	Berries	Cooked
Spruce and birch		Inner bark, spruce needles	Boil inner bark, make tea with needles

Any vegetation north of the tree line is edible, but do not take any of it in large quantities until you know its effect upon your system.

ICELAND MOSS (Cetraria islandica).

Where found - on sandy soil.

Appearance - looks very much like an upright brown seaweed.

POISONOUS PLANTS IN THE ARCTIC (TUNDRA) AND SUBARCTIC

Plants which produce poisoning upon contact do not exist in these cold regions.

The following eight kinds of plants are poisonous only when eaten. They are internal poisons and are the only dangerous ones you are apt to encounter on the tundra and in the subarctic regions adjoining the tundra. You will not find these poisonous plants above the tree line.

BANEBERRY (Actaea)

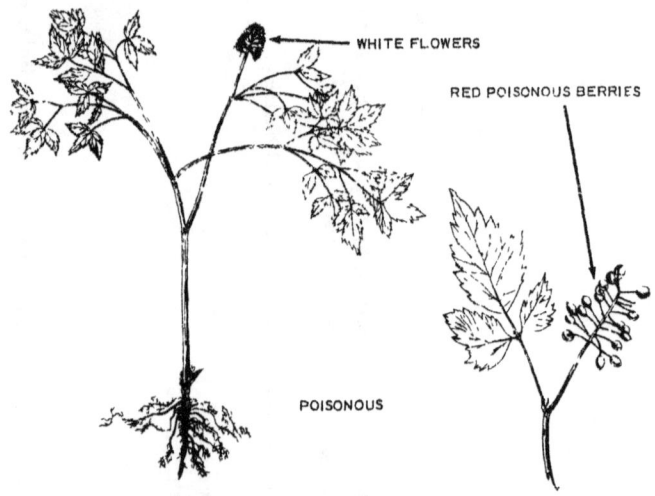

Figure 37. Baneberry (Actaea)

Description.

Perennial from thick rootstock. Stems smooth or somewhat hairy, 2 to 3 1/2 feet high. Leaves large, divided into three leaflets. Leaflets thin, usually lobed and coarsetoothed. Flowers small, white, many in a spikelike cluster at the top of the stem, each flower with 4-10 small, white petals. The fruit is a round, multiseeded berry, red or white. Each berry is attached to the stem by a short, thick stalk, the white-berried plant having red stalks.

Where Found.

Woods and thickets. This is a typical plant of the North Temperate Zone, especially from about latitude 40° N. to the Arctic and subarctic areas of Europe, Asia and North America.

Conditions of Poisoning.

The berries of this plant are poisonous. As few as six berries can cause increased pulse, dizziness, burning in the stomach, and colicky pains. The rootstock is a violent purgative and emetic.

FALSE HELLEBORE (Veratrum)

Description.

Perrenial. Stem stout, erect, 3-8 feet high, rising from a thick rootstalk. Leaves alternate, broadly round-oval with pointed tip, clasping the stem; blade smooth above, hairy beneath; veins parallel. Small flowers in large terminal spikelike clusters with drooping branches; three petals, often greenish, but sometimes white.

False Hellebore (Veratrum)

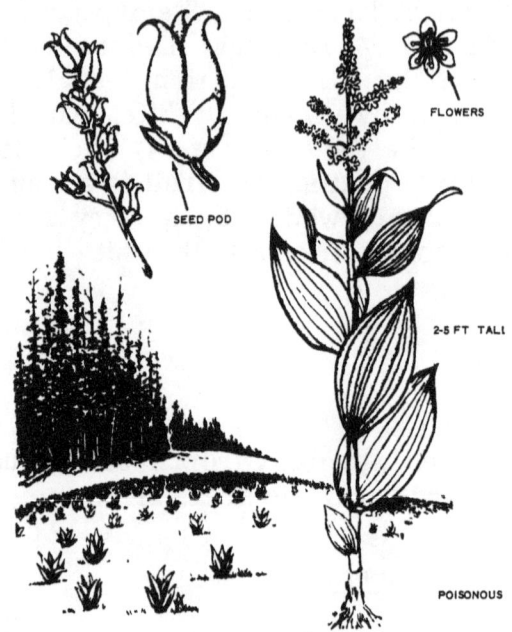

Figure 38. False Hellebore (Veratrum)

Where Found.

Swamps and low grounds in Europe, Asia, and America from about 40° N. latitude. Occurs on the edge of the Arctic, although probably rare on the tundra.

Conditions of Poisoning.

Fatalities among humans from eating false hellebore are rare but are more common to sheep and other animals. Symptoms are salivation, vomiting, purging, abdominal pain, muscular weakness, general paralysis, tremors, spasms, and occasionally convulsions. Death results from asphyxia.

POISON WATERHEMLOCK (Cicuta)

Description.

Perennial. Stems 3 1/2 to 7 feet high, stout, jointed, hollow between the joints, reddish. Leaves alternate, divided into narrow leaflets with toothed edges and the leaf veins end at or near the tooth notches. Leafstalks sheath the stem. Rootstalk short, ringed on the outer surface and often, especially when young, has many fibrous rootlets; when older it has many spindle-shaped roots bunched at the base. When root and lower stem are split lengthwise, many cross-partitions or chambers can be easily noticed. Plant exudes drops of a yellow aromatic oil which gives it a characteristic odor. Flowers small and white, in umbrellalike clusters at the top of the stalk.

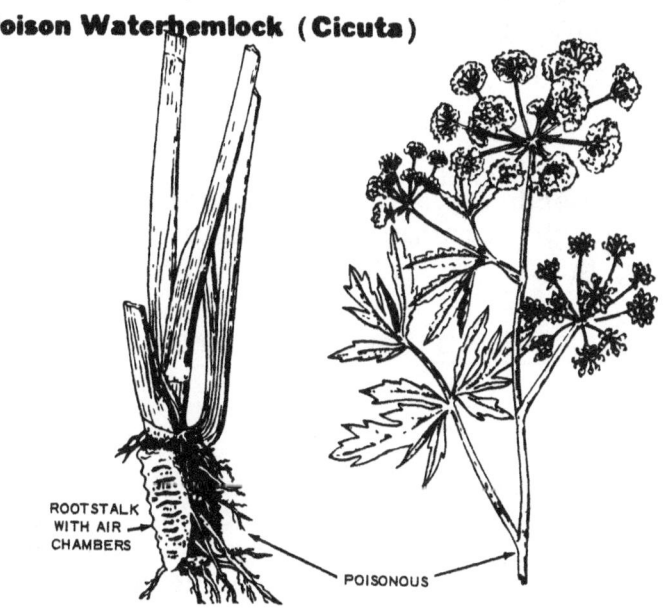

Figure 39. Poison Waterhemlock (Cicuta)

Where Found.

Wet meadows, ditches, along streams, and around tundra lakes. This plant belongs to the parsley, carrot and parsnip family, which contains many well-known edible plants, but it is better to avoid all members of this family as food in northern areas, since the related waterhemlock is fairly common in the North Temperate Zone.

Conditions of Poisoning.

A piece of waterhemlock root about the size of a walnut is said to be sufficient to kill a cow. This plant contains a sticky, resinlike substance called cicutoxin. It is most concentrated in the roots but is present in all parts of the plant. Symptoms are stomach pains, nausea, vomiting, weak and rapid pulse and violent convulsions.

Treatment

In cases of hemlock plant poisoning, make the patient vomit, then give a cathartic. If vomiting is produced promptly, the victim is likely to recover.

VETCH, LOCOWEED (Astragalus)

Description.

Perrenials. Stems erect or spreading. Leaves alternate, each with many leaflets, some smooth, some hairy. Flowers in spikelike clusters at top of stem. Individual flowers pealike in structure with five petals, white, yellow or purplish.

Vetch, Locoweed (Astragalus)

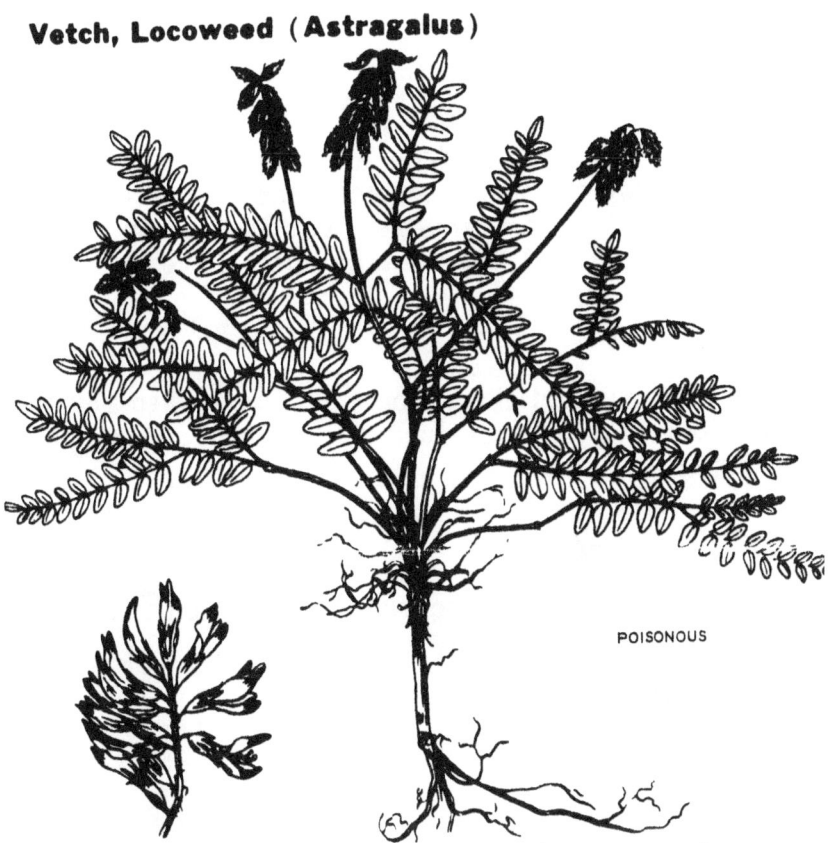

Figure 40. Vetch, Locoweed (Astragalus)

Where Found.

These plants occur rather abundantly in meadows, on hillsides and on tundra throughout the North Temperate Zone.

Conditions of Poisoning.

Several species of locoweed have been reported as toxic. Avoid all kinds to be on the safe side.

BUTTERCUP (Ranunculus)

Description.

Buttercups vary in size from a few inches to nearly 3 feet tall. Most kinds, especially those in Arctic regions, are diminutive in size, with divided or deeply notched leaves. All kinds have yellow flowers.

Where Found.

Widely distributed throughout the North Temperate Zone and well into the tundra of Europe, Asia and America.

Conditions of Poisoning.

If the leaves are eaten, severe inflammation of the intestinal tract may result.

Buttercup (Ranunculus)

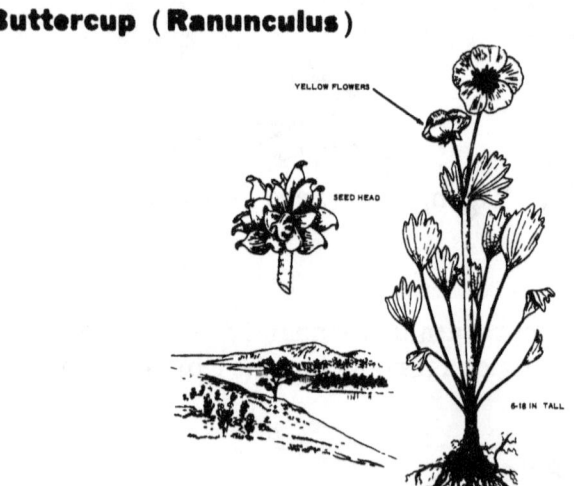

Figure 41. Buttercup (Ranunculus)

MONKSHOOD (Aconitum) and LARKSPUR (Delphinium)

Description.

Monkshood is a perennial, 2-4 feet high. Leaves alternate, the upper clasping the stem, palmately veined, and lobed or divided; leaves and stems somewhat hairy or sticky. Flowers hooded or helmet-shaped, usually blue.

Larkspur is similar to monkshood, but the flowers are not hooded. Most kinds develop two spurs at the base of the flower.

Monkshood (Aconitum) and Larkspur (Delphinium)

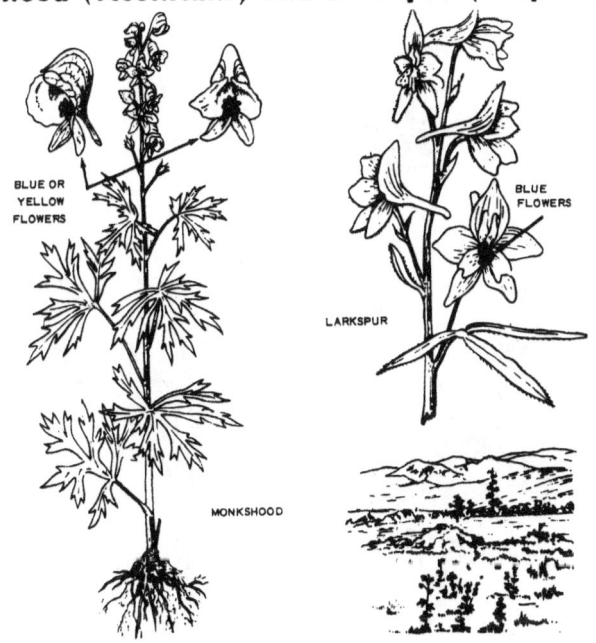

Figure 42. Monkshood (Aconitum) and Larkspur (Delphinium)

Where Found.

Monkshood and larkspur are distributed widely over the North Temperate and subarctic zones, especially in mountainous regions.

Conditions of Poisoning.

This plant, while poisonous at all times, seems to be most poisonous just before flowering. Symptoms are muscular weakness, irregular and labored breathing, weak pulse, bloating, belching, constant attempt at swallowing and pupils contracted or dilated.

DEATH CAMAS

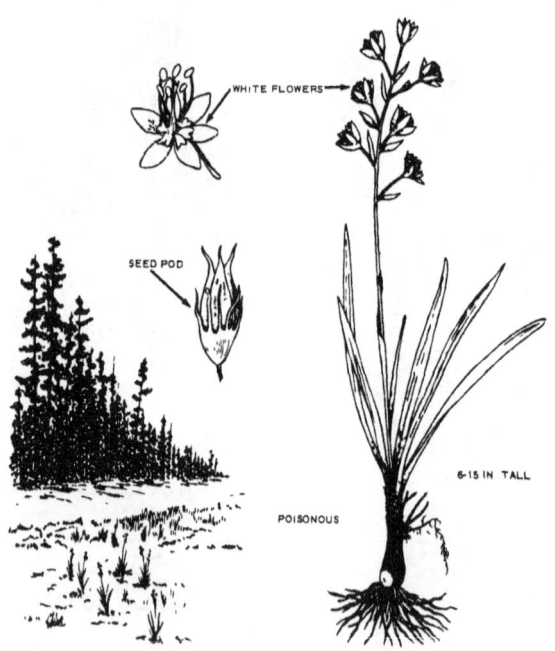

Figure 43. Death Camas

DESCRIPTION. Plant growing from a bulb. Stems 1 to 2 feet high, leafy, with linear leaves clasping the stem at the base. Flowers greenish-white in loose terminal clusters — contrasted with wild onion, which has flowers closely aggregated at the top of the stem; also no onion odor in death camas.

WHERE FOUND. The death camas grow in meadows, and on edges of forests in parts of western subarctic North America and eastern Siberia; also grow farther south to western United States.

CONDITIONS OF POISONING. The death camas contains the toxic alkaloid, zygadenine, in all parts of the plant from the bulb to the seed. Children have been known to be poisoned by eating the bulbs, probably mistaking them for onions.

WATER

In the winter, ice and snow provide water but fuel is needed to melt them. Never waste fuel in melting snows or ice when drinkable water from other sources is available. In the summer there is plenty of water in lakes, streams and ponds. Surface water on the tundra may have a brownish color, but it is drinkable.

Whenever possible, melt ice for water rather than snow - you get more water for the volume with less heat and time. If you melt snow by heating, put in a little snow at a time and compress it - or the pot will burn. If water is available, put a little in the bottom of the pot and add snow gradually.

If the sun is shining, you can melt snow on a dark tarpaulin, signal panel, flat rock or any surface that will absorb the sun's heat. Arrange the surface so that melt-water will drain into a hollow or container.

Use old sea ice for drinking water. It is bluish, has rounded corners when broken, and is free from salt. New sea ice is gray, milky, hard and salty; don't drink it. Water in pools at the edges of ice floes is probably too salty to drink. Icebergs are good sources of fresh water and should be used if they can be approached safely.

If short on fuel, you can melt snow in your bare hands. It's best, however, to fill up on water at mealtime once you have melted ice or snow.

If fuel is plentiful, try to drink at least 2 quarts of hot beverage or water daily instead of cold water or snow.

COOKING

For cooking, the fire should not be too large. If possible, it should be built between two logs or stones on which the cooking utensils may rest. Another method is to suspend the pot over the fire from a pole, the lower end of which is stuck into the ground.

Boiling is the simplest and most practical method of cooking. Long boiling is not necessary. If meat is cut into small pieces and put into cold water it will be cooked sufficiently by the time the water has boiled a couple of minutes. Drink the water in which the meat is cooked.

If snow is used for cooking, place a small amount of it in the pot at first, adding more as it melts. If the pot is crammed full of soft, spongy snow this may act as a blotter, absorbing the first water that melts and allowing the bottom of the pot to burn. This may be avoided by tipping the pot.

If you haven't any pot, the simplest method of cooking meat is to broil thin strips over hot coals, holding them on the forked end of a long stick. Larger pieces can be stuck on a stick and suspended over the fire, turning them from time to time.

CLOTHING AND SHELTERS

Arctic Clothing

The arctic clothing issued to site personnel by the USAF is the best that science can devise. The purpose of the clothing (or any clothing designed to provide warmth) is to keep the wearer surrounded by his own body heat. Arctic clothing is worn in layers, in order to provide an air space between each layer which will help to hold the heat. For this reason, the clothing should always be a loose fit. Too-tight clothing will not leave space for insulating air, and will impair circulation as well. Always be certain, therefore, that your arctic clothing, especially footgear, does not fit too tight. Garments of the same size marked "inner" and "outer" allow for this layer of air.

All RCAS personnel are classified by primary duty assignment into two categories: Indoor and Outdoor. Issue of Arctic clothing to personnel of the Indoor classification will be as follows: wool pile cap; jacket with hood (Parka); outer trousers; mitten set; gauntlet; woolen finger gloves; flying boots; sunglasses and case; suspenders; and aviator's kit bag. Clothing issued to personnel of the Outdoor classification will be as follows: wool pile cap; jacket with hood (Parka); outer trousers; mitten set; gauntlet; woolen finger gloves; sunglasses and case; inner mukluks; two (2) pairs felt liners for mukluks; aviator's kit bag; shoe-paks; insulated vest; and face mask.

Minimize perspiration -- it leads to freezing. When exerting yourself, cut down perspiration by opening your clothes at the neck and wrists and loosening the waist. If you're still warm, remove mittens and headgear, or remove a layer or two of outer clothing. When you stop work, replace clothing to prevent chilling.

Wear loose clothing. Tight clothes cut off circulation and increase danger of freezing. Keep your ears covered with a scarf.

Don't make your boots too tight by wearing too many socks, however, if your boots are big enough, use dry grass or other material for added insulation around your feet. To maintain the value of this insulation you will have to dry the material and fluff it up when it becomes packed.

Felt boots, mukluk boots, or moccasins with the proper socks and insoles are best for dry cold weather. Shoepaks (rubber-bottomed, leather-topped boots) are best for wet weather. The thermo boot can be worn in any weather. Improvise footgear by wrapping parachute cloth or other material lined with dry grass or kapok around your feet.

Keep your clothing as dry as possible. Brush snow from your clothes before you enter a shelter or go near a fire. Beat out frost before warming garments -- dry them on a rack before the fire. Dry socks thoroughly. Don't get boots too near fire.

Wear one or two pairs of wool mittens inside a windproof shell. Avoid removing mittens but if you must remove them, warm your hands inside your clothes -- once they get cold you're in trouble.

In strong wind or extreme cold, wrap yourself in your parachute and get into a shelter or behind a windbreak.

At night, arrange dry spare clothing loosely around your shoulders and hips -- it will help keep you warm.

If you fall in water, roll in dry snow to blot up moisture. Roll, brush off snow and roll again, until all water is absorbed. Do not remove shoes until you are in shelter.

Dirt or grease reduces the insulating value of Arctic clothing, although it does not destroy it as moisture does. It follows that this clothing should be kept clean so that it will insulate most effectively.

Repair Your Clothing

Body heat will escape at an alarming rate from even the smallest openings under conditions of Arctic exposure, as encountered when walking over the Arctic sea ice with the air temperature at 40 or 50 below and a 10 or 15-mile wind.

Be sharp. Carry a couple of strong needles (already threaded) in your cap visor and repair your clothing before you bed down at night.

Shelters and Shelter Living

The choice of a shelter site will depend primarily on the proximity of food and fuel, bearing in mind not to stray too far from your best signal — the aircraft. The shelter should be kept small and suited to the needs of the survivor. A large shelter will be hard to keep warm and will require more effort to build. A palace is not needed — just a place to shield you from the wind and snow, to dry your clothing, and to sleep. There are some excellent building materials in the Arctic, such as evergreen boughs, snow or ice blocks, sod blocks, and/or rocks. A survivor lucky enough to have a parachute has no problem with building material.

Before building starts the shelter should be well planned — it will pay off. Keep it small — this bears repeating for it is important. Locate the entrance crosswind to prevent wind from blowing in or snow from drifting inside. Whenever possible, the shelter should be located near some natural windbreak such as a bank of trees or a snowdrift. Avoid the base of a slope or cliff, however, as a snowslide or rockslide is always a possibility. The shelter should be adequately ventilated. It bears repeating that carbon monoxide poisoning is an ever-present danger in shelter which contains a fire.

In the Summer

You will need shelter against rain and insects. Choose a camp site near water but on high, dry ground if possible. Stay away from thick woods as mosquitoes and flies will make your life miserable. A good camp site is a ridge top, a cold lake shore, or an area that gets a breeze.

If you stay with the aircraft, use it for shelter. Cover openings with netting or parachute cloth to keep out insects. Build your fire a safe distance from the aircraft and do your cooking outside to avoid carbon monoxide poisoning.

A simple outdoor shelter may be made by hanging a tarpaulin or a parachute over the aircraft wing and anchoring the ends to the ground by weighting with stones. A tent can be quickly improvised by placing a rope or pole between two trees or stakes and draping the parachute over it, making the corners fast with stones or pegs. However, a pole tepee or well-constructed lean-to is much more satisfactory.

A fine shelter for drizzly weather and protection against insects is a tepee made from your parachute. In it you can cook, eat, sleep, dress, and make signals without going outdoors. Use 6 panels of parachute for a 2 man shelter, 12 to 14 panels for a 3-man paratepee. Never use more than half the panels of the parachute. The method of constructions is shown in the illustration. This shelter is worth building if you decide to stay in one spot for some time.

Avoid sleeping on the bare ground. Provide some sort of insulation between the ground and yourself -- soft boughs are good. Pick a bed site on level, well-drained ground free from rocks and roots. If you must sleep on bare ground, dig depressions for your hips and shoulders and try out the site before you set up the shelter or spread your bedding.

In The Winter

Snow is one of the best building materials in the Arctic. It is an excellent insulator, is easy to work with, and is plentiful. There are many types of snow shelters from a simple windbreak to a snow house or "igloo". There are two general approaches to a snow shelter: you can dig down into the snow and live in the

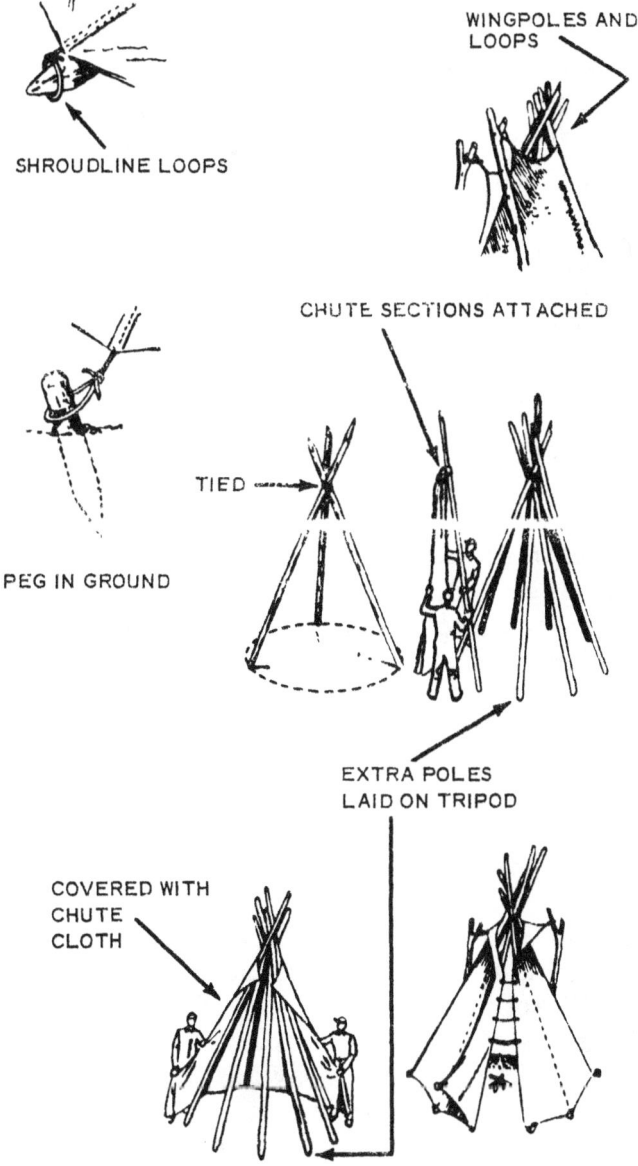

Figure 44. Construction of Paratepee

Figure 45. Paratepee

hole, or you can cut blocks of snow and build upward. The former is easier and requires less skill.

The only equipment needed to construct a snow shelter is a long knife or digging tool. It takes 2 to 3 hours of hard work to dig a snow cave and at least twice as long for the novice to build a snowhouse.

Since they cannot be heated much above freezing, life in snow shelters is rather rugged. It takes several weeks to accustom oneself to the effects of living in such a cold atmosphere. You will require more food (especially fat) and hot drinks.

Regardless of how cold it may get outside, the temperature inside a small well-constructed snow cave probably will not be lower than -10° F. (even when it is -50° F. outside). A small 1-man snowhouse should not get colder than -20° F. Body heat alone will raise the temperature in a snow cave 45° above that of the outside air. A burning candle will raise the temperature 4° and burning Sterno (small size, 2-5/8 oz.) will raise the cave temperature 28°.

When you remove your sleeping bag in the morning, warm air released into the cave will raise the air temperature from 6° to 10°. If it is -50° outside, the temperature of a snow cave (with one man resting and a Sterno stove burning) will be about 32° F., while it may be 40° F. or above in a snowhouse under similar conditions.

While a tent can be erected or struck in a few minutes its weight and bulk pose problems if you are traveling. A tent may feel more comfortable if adequate fuel is available for heating but they are noisy in Arctic storms and it may be hard on the nerves to hear the constant flapping. It is also possible that fire or wind may cause damage or complete loss of the tent.

You will need shelter against the cold. Don't live in the aircraft — it will be too cold. Try to improvise a better insulated shelter outdoors.

Camp in timber if possible, to be near fuel. If you can't find timber, choose a spot protected from wind and drifting snow. Don't camp at the bases of slopes or cliffs where snow may drift heavily or come down in avalanches.

Keep the front openings of all shelters crosswind. A windbreak of snow or ice blocks set close to the shelter is helpful.

In making shelters, remember that snow is a good insulator. Caves in the snow are usually rather satisfactory. However, keep your clothing dry when digging snow caves. Work slowly to avoid excessive sweating.

In timberless country, make a simple snow cave or burrow by digging into the side of a snowdrift and lining the hole with grass, brush, or tarpaulin. Snow caves must be ventilated. If the snow isn't deep enough to support a roof, dig a trench in a drift and roof it with snow blocks, tarpaulin, or other materials.

In wooded country make a tree-pit shelter if snow is deep enough. Enlarge the natural pit around a tree trunk and roof it with any available covering.

Look for cabins and shelter houses. They are likely to be located along bigger streams, at river junctions, along blazed trails in thick, tall timber leeward of hills.

Prevent carbon monoxide poisoning by providing good ventilation for closed shelters in which a fire is burning.

Don't sleep directly on the snow. Provide insulation under your sleeping bag or body. Lay a thick bough bed in shingle-fashion, or use seat cushions, tarpaulins, or even an inverted and inflated rubber life raft if available.

Keep your sleeping bag clean, dry, and fluffed up to give maximum warmth. To dry the bag, turn it inside out, beat out frost, and warm it before the fire — but don't burn it. Turn over with rather than in the sleeping bag.

If the aircraft has been damaged on landing, stay away from the aircraft until the engines have cooled and any spilled fuel has evaporated.

Then check on the condition of the crew and passengers. If bailouts have occurred, look for isolated members of the party. Necessary medical aid should be given, either based on material in the next section of this chapter or on previous Air Force training. Make the injured men comfortable, and be careful in removing casualties from the aircraft, particularly men with injured backs and fractures.

If the weather is severe, throw up a temporary shelter, using parts of the aircraft and parachute canopy. If you need a fire, start one at once. Hot drinks, if available, will be stimulating.

If the emergency radio is operative, use it as soon as possible. If it is not operative, gather all available signaling equipment and use it as outlined in the section on signaling. Sweep the horizon with the signal mirror at frequent intervals. If you use the emergency radio or the aircraft communications system, be careful to conserve your electrical power. Use this equipment according to procedures given in your briefing.

After these basic essentials have been covered, relax and rest until you are over the shock of the crash and the excitement of the unexpected landing. It is most important to regain your equilibrium and to settle your nerves. Leave extensive preparations and planning until later.

After you have rested, organize the crew and your temporary camp. Organization and planning are as important in themselves as they are in the actions that result from them. Case histories of many previous unsuccessful survival situations show that the main pitfall of attempts at survival have been the blind and almost hysterical actions of the men. Intelligent and considered planning is absorbing and gives the mind a discipline and orderliness which are so vital in emergencies such as this.

The first step in the organization of the party will be to appoint individual members to specific jobs, based on previous knowledge of their abilities and their physical and mental capacities to handle them. Pool all food and equipment in charge of one man. Prepare a shelter in a manner indicated later in this chapter, to protect yourself from rain, sun, snow, wind, cold, or insects. Collect all possible fuel, the variety of which will be determined by your geographical location. Try always to have at least a day's supply of fuel on hand. Then look for water and for animal and plant food as prescribed in various specific discussions throughout this manual.

As another start toward organization and self-discipline, begin a log book, making brief but accurate entries including:

Date and cause of crash.

Probable location.

Roster and condition of personnel.

Inventory of supplies.

Weather conditions.

Other pertinent data.

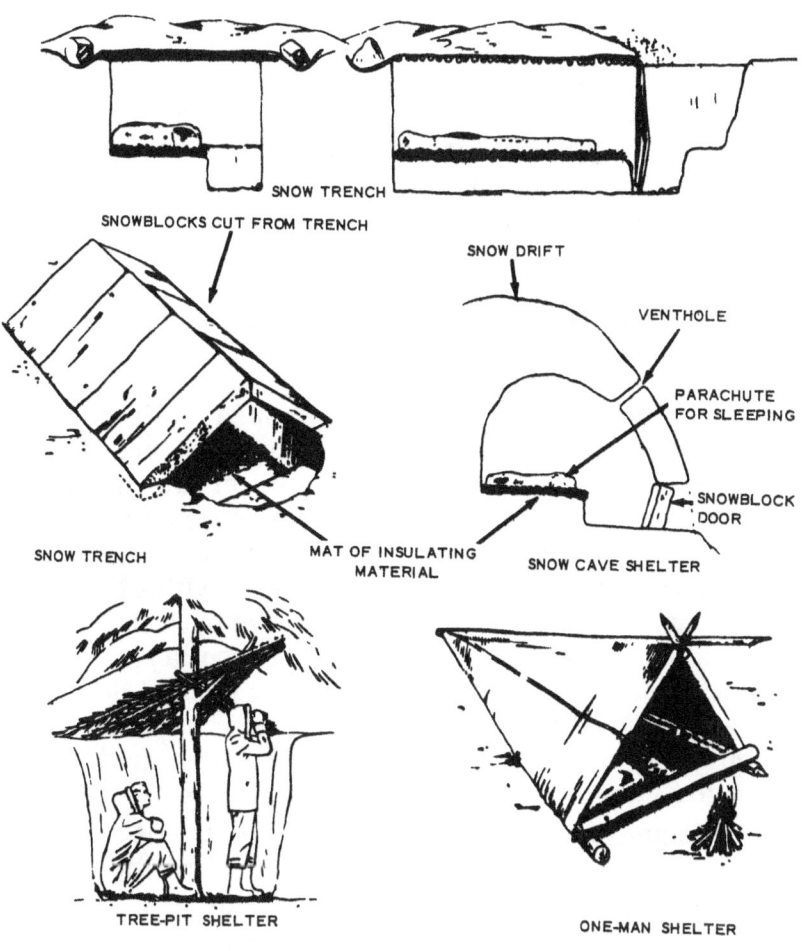

Figure 46. Easy to Construct Shelters

In fact and fiction, countless survival stories have included references to patiently kept log books. The log book is valuable both in that it gives the survivors a record of their activities and the passage of time, and in that it furnishes military authorities with valuable information for future survival training planning.

Determine your position by the best means available (see section on direction and position finding), and include this position in any radio messages you may transmit. If position is based on celestial observations, transmit the observation also.

Snowhouse or "Igloo"

An excellent shelter is the snowhouse or "igloo" which requires more effort and skill to construct than the other types. Such an undertaking should not be attempted by one or two men but is more suitable for a group of survivors. Snow of the right consistency is the first prerequisite. Test the snow with a blunt stick seeking a steady response to pressure. If the stick jerks as it passes through crystallized layers it is not suitable. When snow of the proper consistency is located draw a circle the desired size of the snowhouse on the surface, and criss-cross it with a grid laid out in the size of blocks. This will be used as a sawing guide and the blocks will be cut from inside the snowhouse. In this way, standing room can be obtained without building a six-foot roof. The best block size is about three feet by one and one-half feet by half a foot (3 x 1 1/2 x 1/2). Saw out the first block, then dig a hole next to it and make the undercut. After cutting out the first block you can stand in the hole and make the blocks more easily. Build the circle in a spiral manner, leaning the block inward to obtain the desired ceiling height. The last block will be difficult and must be made by the cut-and-fit method. Shave the roof smooth, and throw snow over the outside to chink cracks, then heat the shelter until the walls have the consistency of slush. The heat should be removed at this time, and the inside walls allowed to freeze to a glaze. The snowhouse is now finished. Keep the bed platform up in the warmer air, and do not sleep directly on the snow. Ventilate the roof with a small hole and if possible, use some sort of non-metallic tube so the escaping warm air will not enlarge the hole. The useful life of a snow-house is only three or four weeks, after which it turns to ice and cannot be kept warm. Another must be built when this happens, but this can be done with the increased skill that comes with practice.

Snow Caves

A snow cave can be dug with a shovel, a knife, or even a ration tin.

Dress lightly while digging, it is hard work and you may easily become overheated which will cause your clothing to dampen with perspiration.

First, dig a tunnel sloping upward, then level off and cut an entry way. About 18 inches above the entry way, build the bed platform. To conserve heat this may be built just large enough to sleep, dress and undress while lying in the sleeping bag. If desired to conserve heat, the sleeping shelf may be walled in. In addition to a ventilation hole through the roof, there should be another at the door.

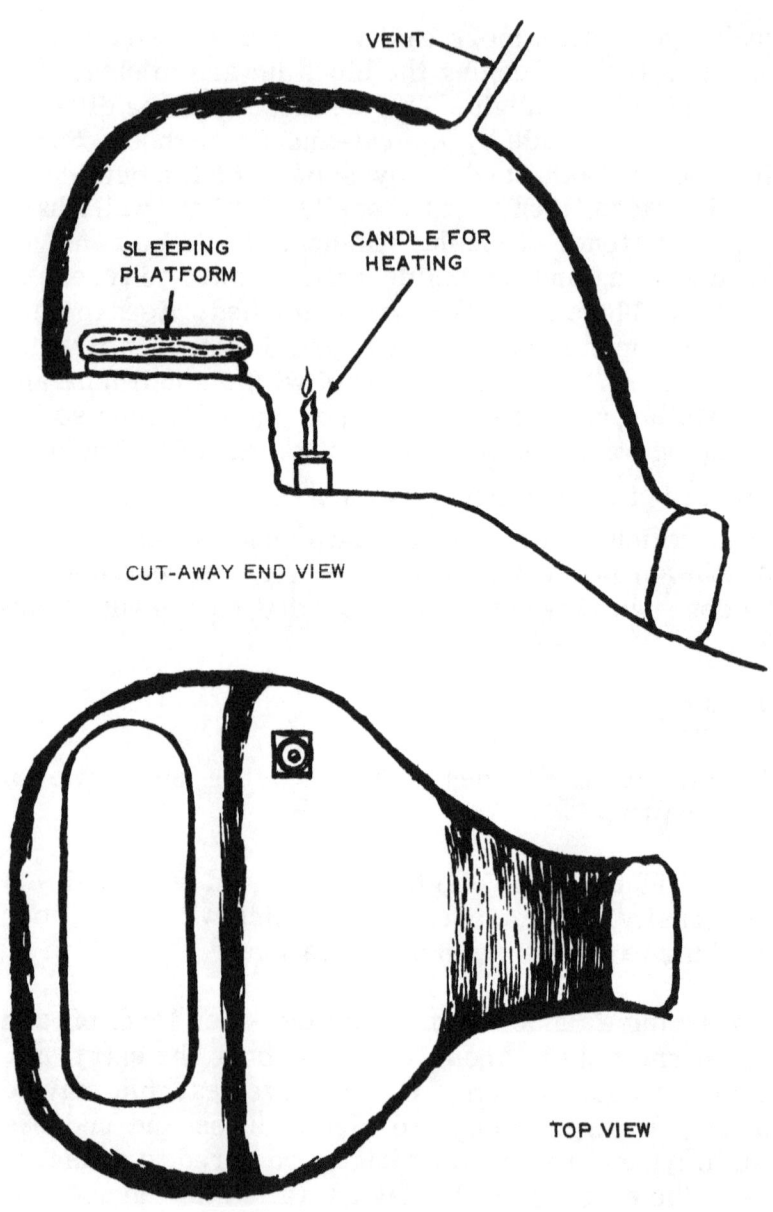

Figure 47. Snow Cave

Take a shovel in with you at night in the event you need to dig yourself out. One storm may deposit a great quantity of snow, which may be very hard to remove without a cutting tool.

Always check ventilation holes before lighting a stove in a snow shelter. Carbon monoxide is a great danger.

Going In and Out

Keep entrances and exits to a minimum to conserve heat. Fuel is generally scarce in the Arctic and it is important to keep the door closed as much as possible.

If someone must go out, plan to have him do as many necessary activities as possible, such as gathering fuel, snow or ice for the cookpot, emptying the trash, etc. To expedite matters, the trash box may be kept just inside the door, and equipment may be stored in the entry way. Necessities which cannot be stored inside should be kept just outside the door.

Once the door is sealed for the night, you will make yourself very unpopular with your companions if you keep running back and forth. It can also be hard on you.

The Snow-Brush is a Lifesaver

Always brush your clothes before entering the shelter. Once inside a warm shelter any snow remaining on the clothing will melt and when the clothing is again taken outside, the water formed will turn to ice and reduce the insulation value. Under living conditions where drying clothes is difficult, it is easier to keep clothing from getting wet than to dry it out later.

If you cannot get all of the snow from your outer clothing, take it off and store it in the entry way or on the floor far from any source of heat so it will remain cold. If ice does form on clothing, it should be scraped off with a knife or beaten out with a stick.

Cooking Discipline

In the cramped quarters of any small emergency shelter, pots of food or drink can easily be kicked over accidentally. In a survival situation, this may seriously affect your chances of coming through safely. The cooking area, even if it is only a Sterno stove, should be located out of the way, possibly in a snow alcove.

It is generally safer to fill gasoline stoves outdoors or in a separate room in the shelter. A period of good weather should be chosen for this task. At this time, any needed maintenance work may be done. In so doing, there is less likelihood that the stove may burn out or clog up while you are using it.

Ice is better than snow for providing water, and the compact snow at the base of a drift is better than the powder snow on top.

Living in Your Sleeping Bag

With little else between you and the ice, your sleeping bag will be very valuable. **Keep it dry.**

It will be more pleasant if you do most of your undressing and dressing in the sack. It's a good idea to have a hot snack or a hot drink just prior to getting into the sleeping bag. You will find it more comfortable to sleep in long woolen underwear (stored in the bag for that purpose) rather than in the clothes you wore during the day. Also put on dry bed socks. Outer clothing makes good mattress material. A parka held over the stove for a moment to absorb some heat makes a good footbag. Shirt and inner trousers may be rolled up for a pillow. Socks and insoles should be separated and dried at the top of the shelter when there is not a pot on the fire throwing off steam. Drying may be completed in the sleeping bag by stowing around the warmest section of the body, the hips.

To keep frost (from your breath) from wetting the sleeping bag, improvise a frost cloth from a piece of clothing, a towel, or some parachute fabric. Wrap this around your head in such a way that the breath is exhaled through a sort of tunnel opening in the fabric leading to the outside of the bag. It is easier to dry a piece of fabric than the sleeping bag. If you should get cold during the night, exercise by moving your feet up and down. Fluff out the feathers in your bag by beating against the inside of it with your hands, try eating something, or make yourself a hot drink.

Don't forget your tin can urinal. One experience getting up to go outside at -50° F. will cure you.

In a snow cave, getting up is a slightly slower process than going to bed. Take your time! Begin dressing — inside the bag.

Toilet

It is standard practice in snow shelter living, to relieve

yourself indoors whenever possible. This practice conserves body heat. If the snowdrift is sufficiently large to dig connecting snow caves, one may be used as a toilet room. If not, tin cans may be used for urinals, which should be emptied in a remote corner opposite to the side from which snow is taken for cooking. Try to have your bowel movement just prior to leaving the shelter in the morning and remove the fecal matter with the trash.

Beware of Drifting Snow

Discipline should be exercised so that "there is a place for everything and everything is in its place". A tent or snow cave can be buried so deep by drifting snow that you may have difficulty finding it, if it isn't marked with a long pole. Therefore, it is a certainty you can't leave small objects lying out on the snow and expect to find them again. A little organization should minimize the chances of losing equipment.

Axes and Knives

You can make yourself a wooden snow knife or saw if you can find a piece of wood 20 inches long and 3 inches in diameter. Split a plank, then shape a handle on one end, and you have a knife — cut teeth in the edge and you have a snow saw. You can increase the efficiency of both items by wetting them once or twice to glaze the surface.

Use your tools carefully. Don't pry snow blocks roughly with the saw, it may snap. Frozen green wood is like a rock and chopping it with an ax is dangerous. The ax may be chipped or it may strike a glancing blow that could injure someone.

Firearms

During the winter remove all lubricants and rust preventive compounds from your weapons, strip them completely, and clean all parts. Use gasoline or lighter fluid, if available, or boiling water. Normal lubricants thicken in cold weather and slow down the action. In cold weather, weapons function best when absolutely dry.

A major problem is to keep snow and ice out of the working parts, sights, and barrel. Even a small amount of ice or snow may disable your weapon, improvise muzzle and breech covers, and use them. Carry a small stick in your pocket to clean the sights and breech block.

Weapons sweat when they are brought from extreme cold into a heated shelter, when they are taken out again into the cold the film of condensation freezes. This ice may seriously affect their operations, leave firearms outdoors or store them in unheated shelters.

If your shelter is not much warmer than the outside temperature you may bring your weapon inside, but place it at or near floor level where the temperature is lowest. When you take it into a heated shelter the weapon may sweat for an hour, wait until it is warm before drying and cleaning.

If a part becomes frozen, do not force it abruptly, warm it slightly if possible, and move it gradually until unfrozen. If it cannot be warmed, try to remove all visible ice or snow and move it gradually until action is restored.

Before loading your weapon, always move the action back and forth a few times to insure that it is free.

If your weapon has a metal stock, pad it with tape or cloth, or pull a sock over it to protect your cheek. If your hand becomes frozen to the metal parts, don't try to pull it away, urinate on the skin where it is stuck to the metal.

Firemaking

A fire is essential in any survival situation, it performs many tasks besides keeping the survivor warm. It can be used to melt snow or ice for drinking, to cook food, to dry clothing, or to signal a rescue aircraft.

IMPROVISED STOVE USING AVIATION FUEL

IMPROVISED STOVE USING WICK TO
BURN OIL OR ANIMAL FAT

Figure 48. Improvised Stove

Building a good fire starts before the match is lit. There are certain prerequisites which must be considered before the survivor can build his fire. These are: base, windbreak, source of fuel, and of course, source of flame.

STRIKING KNIFE ON BOTTOM OF/WATERPROOF MATCH BOX

Figure 49   Waterproof Match Box

A base for the fire will be needed so that it will not melt into the snow and go out. Any non-burnable will serve this purpose. Green logs, flat rocks, sod or metal from the aircraft are examples of base materials. Select a spot that is sheltered from the wind, but do not build it under a snow-laden tree the snow may fall and put out the fire. A windbreak is usually needed to keep the fire from being extinguished or spread by the wind and will also serve the purpose of reflecting the heat into a shelter. The windbreak should also be made of a non-burnable, although snow blocks may be used if they are not placed too close to the fire.

Fuel will be the next problem. Here again, the man who has stayed with the aircraft is fortunate. Many items on the plane can be used for fuel: gas and oil, upholstery, packing materials, and packing containers. The oil must be drained from the engines quickly or it will freeze inside and stay there. Let the oil run out right onto the ground, it will freeze into a soft mat and can then be rolled like tar. It is valuable for signal fires, as it gives off a black smoke against the snow. Do not attempt to burn raw gasoline or throw it on a burning fire -- it is <u>dangerous</u>. A <u>few</u> drops may be sprinkled on tinder to make it easier to start, however, to use gasoline as a fuel, special techniques are necessary. Oil should be mixed with gasoline to make it a safer fuel. If it is necessary to burn pure gasoline, a sand stove is used. Punch holes in the bottom of a can and fill it part way with sand, saturate the sand with gasoline and it will burn for some time. This is the only way to burn pure gasoline safely.

If the man is not with the aircraft or if it was destroyed by fire, fuel must be obtained from nature. In many areas this will be difficult, not so difficult in others. In forested areas, wood is the best source of fuel. Try to find deadwood. Frozen green wood gives off a ringing sound when struck with a hatchet, frozen deadwood will rattle. Driftwood, as well as coal scraped from the sea bottom by the ice, can be found in seacoast areas. There are also deposits of pitch on the seacoasts. Pitch or coal, however, require an extremely hot fire to ignite and this may pose a problem. On the tundra, there are plants which will burn. Willow branches may be used for fuel, and Cassiope plants may be dug from under the snow and burned wet, due to their high pitch concentration. The Cassiope will give off much smoke, but still provides some heat. On the sea-ice the only fuel that can be found from nature is blubber. Of course, to obtain blubber you must first kill a seal. Assuming the survivor has a supply of blubber, it can be burned in the following manner: Make a pyramid of seal bones, and squeeze some blubber onto a small piece of cloth. Place the blubber,

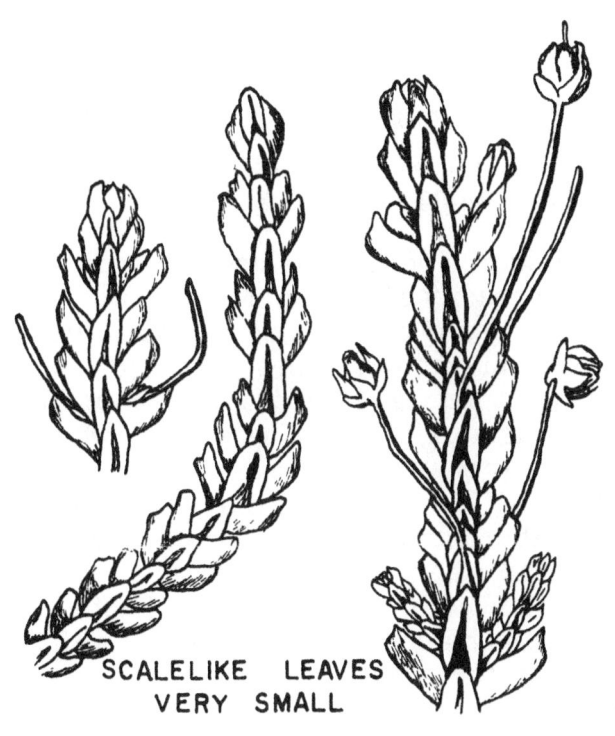

SCALELIKE LEAVES
VERY SMALL

Figure 50. Cassiope

cut in strips, over the pyramid and light the wick underneath. The heat from the wick will cause blubber oil to run down the bones and keep the wick supplied. The smoke will be oily black and have an offensive odor, but intense heat also will be generated. Another method of burning blubber is to place a square of it hairside down on the ice, put two or three ignition tablets from the survival kit on the blubber and light the tablets. The blubber will catch and burn with an intense heat. A square foot of blubber will burn for about three hours in this manner. It is always wise to think twice however, about using food source as a material for fuel. A given amount of blubber will supply the body much more heat if eaten, rather than burned in a fire. The same goes for the hide of a large animal — it will burn, but could possibly be used to much greater advantage as clothing or bedding.

Finally, a source of flame is needed, and this may or may not be readily available. Matches are, naturally, the easiest method of obtaining a flame, although Eskimos who can obtain matches prefer to use flint and steel. If your matches have become wet or lost, there are other methods of obtaining a flame which will be discussed. An easy way to make waterproofed matches is to pour melted wax into a penny box of matches and tear away the container when lf hardess. If no matches are to be had, the next best thing is flint and steel. Flint is furnished in the survival kit (fastened to bottom of match container). Scrape the flint with the blade of a pen knife (or other steel) with a downward motion, directing the sparks at the base of the kindling. If no flint is available, look for some kind of rock from which sparks can be struck. If the rock scars or chips, try another kind. A fire may also be started by firearms — remove the projectile, wad the cartridge, and fire into the base of the kindling which should be in some kind of a pocket so the muzzle blast will not blow it away. A flare fired into a pile of tinder and kindling will also start a fire. In either of these methods, the survivor must consider first whether he can spare the ammunition or the signal. The lens from glasses, a watch crystal, or a camera lens may be used to start a fire but this method requires patience and a steady hand. Focus on a spot on the tinder and wait — it may take an hour or more. Rather than destroy a good camera, merely open the back, set it for time exposure, and use the lens in this manner. Another source of flame is the battery from the aircraft. Here again, a decision is necessary. If the radio is operable, make your fire by other means and save the battery. If not, a wire shorted across the terminals will glow red and start your fire, or sparks may be struck from a terminal with a wire connected to the other terminal. This must be done soon after landing, as the liquid soon freezes rendering the battery useless. As a last resort, a bow-drill may be used to start a fire, but it must be remembered that this method is arduous and sometimes unsuccessful. The right wood is needed, as well as good tinder and infinite patience. The hardwood dowel is pushed down into the softwood board with a block held in the operator's hand. The bow is then drawn back and forth, spinning the dowel and heating the tinder by friction. Again we must state that this method is for the experts, and should be avoided unless all else fails.

Tinder will be needed to get a fire started. Some sources of tinder that may be readily available to the survivor are dry grass, lint, gauze, tobacco, moss, pin feathers, animal nests, wood dust, and packing materials.

Build the fire outside the shelter if possible. More survivors and arctic explorers have been killed by carbon monoxide poisoning from a fire in an unventilated shelter than any other cause. A good windbreak will reflect heat into the shelter and this danger will be avoided. Keep the fire small — a large fire loses much of its heat to the wind and wastes fuel. Much more benefit will result from several small fires ringed around the survivor. Keep the fire going—matches, fuel, and time are wasted getting a fire started after it has been allowed to die out.

Useful Hints

Don't waste your matches by trying to light poorly prepared kindling. Don't use matches for lighting cigarettes — get a light from your fire or use a burning lens. Don't build unnecessary fires — save your fuel. Practice primitive methods of making fires <u>before</u> all of your matches are gone.

Carry some dry tinder in a waterproof container and expose it to the sun on dry days. Adding a little powdered charcoal will improve it. Collect good tinder whenever you find it, cotton cloth is good, especially if scorched or charred. It works well with burning glass or flint and steel.

Collect kindling along trail before you make camp. Keep firewood dry under shelter. Dry damp wood near fire so you can use it later. Save some of your best kindling and fuel for quick fire-making in the morning.

To split logs, whittle hardwood wedges and drive them into cracks in the log with a rock or club, split wood burns more easily. Never swing an ax <u>toward</u> your foot or any part of your body.

To make a fire last overnight, place large logs over it so that the fire will burn into the heart of the logs. When a good bed of coals has formed cover it lightly, first with ashes and then dry earth. In the morning the fire will still be smoldering.

Fire can be carried from one place to another by a lighted punk, a smoldering coconut husk, or slow-burning coals. When you want a new fire, add tinder or a small amount of fuel and fan or blow the smoldering material into flame.

Don't waste fire making materials. Use only what is necessary to start a fire and to keep it going for the purpose needed. When you leave the camp site put out the fire with water and mineral soil. Mix it well until you can insert your hand.

HEALTH HAZARDS IN THE ARCTIC

With the exceptions of finding food and keeping warm, you will find that you will have relatively few health problems in the Arctic. There are no disease carrying insects or dangerous parasites in the far North, and no water is contaminated.

Several important points should be remembered in any Arctic survival situation. Most important of these is to keep your clothing dry at all times. If clothing becomes wet, even if from perspiration, it must be dried immediately. Another important precaution is to avoid becoming overexhausted. Contrary to some popular conceptions, you should take your sleep when you need it. If you should become cold enough whereby the temperature of your body is effected, you will awaken even from a normal sound sleep. It is much the same as when you awaken from the cold at home, and get up to close the window. The big danger is in allowing yourself to become completely overexhausted before taking sleep; you may freeze before you wake up when you finally go to sleep. You should, if possible, avoid any activity which requires a lot of effort. Not only does such effort use up body fuel rapidly, but it may cause perspiration, which will dampen the clothing.

Dig latrines downwind and downhill from the camp. Mess gear should be thoroughly cleaned after every meal.

A well-maintained program of personal hygiene, as

much as the elements will permit, will cut down the chances of disease and sickness, and so enhance your chances of survival over a longer period of time.

Although little trouble will be encountered in keeping food in cold climates, it should still be wrapped in cloth and buried, especially in summer, to prevent insect infestation and spoilage.

Dysentery may result from poor housekeeping in a survival camp, and the remedy for this condition is to sterilize all eating and cooking utensils until the condition disappears.

FREEZING

You will freeze only if the air is carrying away more heat than your body can generate. If you prevent the air from reaching your skin by wearing proper clothing and keep your body in heat balance, you won't freeze. Your whole body must be kept warm to maintain circulation to your hands and feet. Excessive loss of heat from any part of your body restricts circulation, leaving your extremities with little heat. With no heat coming in, your hands and feet are liable to frostbite.

There are four general rules of clothing care which, if followed, will help insure against freezing. These rules can be remembered easily by writing out the word "COLD", and the following rules: "C" - keep clothing Clean; "O" - avoid Overheating; "L" - wear clothing Loose and in Layers; and "D" - keep clothing Dry.

FROSTBITE

Frostbite is the freezing of some part of the body. It

is a constant hazard in subzero operations, especially when the wind is strong. As a rule, the first sensation of frostbite is numbness rather than pain. You can see the effect of frostbite (a grayish or yellow-white spot on the skin) before you can feel it.

Use the buddy system. Watch your buddy's face to see if any frozen spots show; and have him watch yours.

Get the frostbite casualty into a heated shelter, if possible.

Warm the frozen part rapidly. Frozen parts should be thawed in water until soft, even though the treatment is painful. This treatment is most effective when the water is exactly 105°F., but water either slightly cooler or warmer can be used. If warm water is not available, wrap the frozen part in blankets or clothing and apply improvised heat packs.

Use body heat to aid in thawing. Hold a bare, warm palm against frostbitten ears or parts of the face. Grasp a frostbitten wrist with a warm, bare hand. Hold frostbitten hands against the chest, under the armpits, or between the legs at the groin. Hold a frostbitten foot against a companion's stomach or between his thighs.

When frostbite is accompanied by breaks in the skin, apply sterile dressing. Do not use strong antiseptics such as tincture of iodine. Do not use powered sulfa drugs in the wound.

Never forcibly remove frozen shoes and mittens. Place in lukewarm water until soft and then remove gently.

Never rub frostbite. You may tear frozen tissues and cause further tissue damage. Never apply snow or ice; that just increases the cold injury. For the same reason, never soak frozen limbs in kerosene or oil.

Do not try to thaw a frozen part by exercising. Exercise of frozen parts will increase tissue damage and is likely to break the skin. Do not stand or walk on frozen feet. You will only cause tissue damage. Above all, when treating for frostbite make sure not to burn the skin.

FLIES

In northern regions the word "flies" has a special meaning. It refers to the insect pests that come after you in savage hordes - usually two or three kinds at once. They include mosquitoes, black flies, deerflies, and midges. All of them bite; they do not sting. They do not carry disease in their bite. In the far north they come out during the day which may be 24 hours long; in southern regions, mosquitoes prefer the evening. Always on the lookout for a landing place, they swarm about their victim, centering their attack upon exposed skin or places where hair is thinnest. There they alight, probe, and bite away until a hand slaps them, a strong wind blows them away, or a chill rain disposes of them. Only with the end of summer does their onslaught stop.

Protective measures against these flies include head nets, which must stand out from the face so that they do not touch the skin; gloves; and clothing in several layers, for they do not often succeed in biting through two layers of cloth. Tuck your pants into the tops of your shoes or socks. Light-colored clothing will protect you better than dark. Make your bed fly-proof, i your shelter is not. Be sure that your mosquito net is held away from your face and body. A net covering your whole sleepingbag is more satisfactory than one covering just your head.

A smudge fire properly made will furnish relief from flies during a meal or will drive them away from a tent. For a good smudge, clear away all debris and humus down to the mineral soil or permafrost, to prevent ground fire. Build up a brisk blaze of deadwood; let it burn until a bed of coals is formed. Meanwhile, gather a supply of additional fuel as well as a mass of green ferns, leaves, twigs, damp leaf mold, and rotten wood. Place new dry wood on the coals and let it burn brightly; then cover the whole fire with part of the green material. The dense smoke that now rises will banish black flies instantly and will repel most of the mosquitoes. The smoke will also irritate your eyes, so get down close to the ground. This smoke is the lesser of two evils.

Very useful is a bucket smudge built in a pail or pot - it can be moved easily in case the wind changes or can be taken inside your shelter. Put about 2 inches of noncombustible material into the pail before you start your fire.

Soaking in water is the easiest and best all-around treatment for fly bites.

CARBON MONOXIDE POISONING

Carbon monoxide poisoning is a great danger in the Arctic. It is colorless and odorless. Prevention of carbon monoxide poisoning is easy. Maintain good ventilation when a fire is burning. Never fall asleep without turning out a stove or lamp. Don't leave a shelter for long if a stove or lamp is burning and others are immobilized or asleep therein. Always make sure that the entrance is clear of snow and free from all obstacles which might prevent quick exit.

Carbon monoxide burns with a blue flame and is freely generated by a yellow flame. Therefore, when you see a yellow flame, check your ventilation. If you are cooking, lift your pot from the flame; if the flame turns blue, your stove is operating correctly. If possible, hang your cooking pot over the stove so that the bottom of the pot is approximately 3 inches from the top of the burner. Then the flame will stay blue as long as the stove is operating properly.

SKIN CARE

A warm wet pack with no antiseptic added is one of the most effective methods of treating any skin irritation which might become infected. Use warm water and apply with cotton wicking or moistened and warmed moss.

The importance of removing accumulated body oils and perspiration from the surface of the skin is even greater under survival conditions than under normal ones. All such waste material acts as a conductor and drains your body's supply of heat. In addition, you will get such a boost in morale from a bath. The requirements for an Arctic bath are a rag at least a foot square (cotton or wool is best), a cooking pot half full of hot water, and some soap if you have it. (If you have no cloth, use wadded sphagnum moss). Remove as many garments as you comfortably can and loosen the rest. Immerse the washrag, wringing it over the pot so as not to lose water, and start washing under your clothing. You must rinse the rag frequently.

Beards are particularly undesirable in the Arctic, for frost collects on them and cannot be readily removed. So if you can't shave, trim your beard as short as you can. A pair of small scissors would be a useful addition to your survival equipment.

If you come into contact with natives, watch to see if they scratch. If they do, they are probably infested with lice. Avoid bodily contact with them.

SNOW BLINDNESS

Snowblindness results from failure to wear sunglasses in the constant glare of the sun and the snow. It is actually a sunburn of the thin film of skin covering the eyes. The affected person will not be able to see, and his eyes will sting and water. Sight will be restored and his condition will disappear after two weeks, and herein lurks the real danger of this condition - the survivor will be essentially helpless until his sight is restored. This, obviously is very serious if the man is alone, or has only one or two companions. He will be a burden on the rest of the party for the entire time. It makes good sense, therefore, to wear sunglasses at ALL times in snow country - even on an overcast day.

This action will preclude the possibility of snowblindness. Wet tea bags applied to the eyes will help to alleviate the pain, but will not restore the sight any more rapidly.

Actually snowblindness can occur on cloudy days as well. Prevention is the best cure. A handy substitute for goggles or sunglasses is a piece of wood, leather or other material, with narrow eye slits cut in it.

This eyeshade is good in a blizzard because you can brush off the slits to keep them clean, while glasses may become frosted over.

Treat snowblindness by protecting the eyes from light and relieving the pain. Protect the eyes by staying in a dark shelter or by wearing a lightproof bandage. Relieve the pain by putting cold compresses on the eyes if there is no danger of freezing, and by taking aspirin. Use no eye drops or ointment. Most cases will recover within 18 hours without medical treatment. The first attack of snowblindness makes the victim more susceptible to later attacks.

SUNBURN

You can get badly sunburned in northern regions, even on foggy or light overcast days. Cover up in bright sunlight. Use sunburn ointment or sea marker dye if you have it.

ILLNESS

The prevention of illness is a factor in survival which deserves constant attention. Personnel incapacitated by illness are just as much casualties as severely injured ones. Food must be cooked, water boiled, and waste properly disposed to maintain good health.

For colds, fevers, and intestinal disorders, rest and hot liquids are the best treatments you will probably have available. APC's and aspirin will aid in the control of fever and the relief of pain.

IMPROVISED MEDICAL EQUIPMENT

Bandages and dressings can be made fairly sterile by

boiling or steaming (preferably) them in a covered container. Freezing or charring with heat is also advantageous.

A good substitute for burn ointment can be made by keeping pressure dressing moist with boiled water to which salt has been added (if available).

If regular medical supplies are not available, use your ingenuity and improvise substitutes. Try to keep all equipment as sterile as possible. Extreme heat is your best method of doing this.

SEVERE CHILLING

If you are totally immersed in cold water for even a few minutes, your body temperature will drop. Long exposures to severe dry cold on land can also lower your body temperature. The only remedy for this severe chilling is to warm the entire body. Warm by any means available. The preferred treatment is warming in a hot bath. Severe chilling may be accompanied by shock. Warm yourself from the inside out by drinking hot coffee, hot soup, or plain water. Build a circle of small fires and sit inside the circle. Exercise as much as possible without sweating.

IMMERSION FOOT (TRENCH FOOT)

Immersion foot is a cold injury resulting from prolonged exposure to moisture at temperatures just above freezing. In the early stages of immersion foot, your feet and toes are pale and feel cold, numb and stiff. Walking becomes difficult. If you do not take preventive action at this stage, your feet will swell and become very painful. In extreme cases of immersion foot, your flesh dies, and amputation of the foot or of the leg may be necessary.

Because the early stages are not very painful, you must be constantly alert to prevent the development of immersion foot. To prevent this condition:

Keep your feet dry by wearing waterproof footgear and keeping your shelter dry.

Clean and dry your socks and shoes at every opportunity.

Dry your feet as soon as possible after getting them wet.

Warm them with your hands, apply foot powder, and put on dry socks.

When you must wear wet socks and shoes, exercise your feet continually by wiggling your toes and bending your ankles. When sleeping in a sitting position, warm your feet, put on dry socks, and elevate your legs as high as possible. Do not wear tight shoes.

Treat immersion foot by keeping the affected part dry and warm. If possible, keep the foot and leg in a horizontal position to increase circulation.

TULAREMIA

Tularemia is a disease of rodents, particularly of hares, rabbits, and squirrels. You can catch it from ticks or by handling infected animals, eating partially cooked and infected animals, and handling or drinking infected water. Avoid all rodents that are not active and healthy. Use gloves to skin rodents and discard the hide.

PREVENTING INFECTION

In administering to a wounded person, cut or tear away clothing necessary to get at a wound. Don't touch a wound with fingers or dirty objects. Don't suck any wounds other than snake bites.

Apply sterile dressing to wound with a firm pressure. Tie firmly but not too tightly.

Keep wounded part at rest.

Iodine may be used to sterilize skin areas surrounding a wound but should not be poured directly into an open wound. Let iodine dry in air before applying bandage.

FIRST AID

Illness and injury are always potential survival companions. Situations may arise where treatment is mandatory. But, remember that unless administered by a specially trained person, medical treatment other than the simplest first aid may be dangerous. If you don't know the procedure for treatment or if you have reason to expect medical assistance soon, it is better to merely make the person as comfortable as possible than to injure him more severely by improper treatment. However, the information given in this section will cover most of the injuries that you will encounter in an aircraft accident.

The most likely injuries will be cuts and bruises, fractures, concussions, internal injuries and burns. Keep the injured men lying down, and if they are unconscious or have head injuries, keep them face down. Handle injured men very carefully, especially if they seem to be suffering from a fracture or a back injury. If the men are suffering from head injuries, difficult breathing, flushed face, or bleeding from ears, nose or throat, keep their heads slightly elevated. Give as much treatment for shock as possible under the circumstances if such a condition is noticed.

The following procedures should be followed carefully.

Burns

Don't touch a burned part with fingers. Cover freely with burn ointment. Apply thick gauze pack; bandage firmly. Don't change bandage. If pain is severe, give a morphine injection. Give large amounts of fluid. Keep the burned part at rest. Splints may sometimes be used to good advantage. If necessary to open blisters, use sterilized needle to pierce through the skin at the base of blister. Apply a sterile bandage after draining.

Bleeding

Serious bleeding must be controlled as soon as possible. The following methods, in the order of presentation, are recommended.

Place sterile pad directly on wound and apply pressure by hand or by bandaging firmly. Do not be impatient. Stick to this method, if possible.

Finger pressure on the pressure points shown in illustrations will control arterial bleeding. Arterial bleeding may be recognized by noting any bright red blood spurting freely from a wound.

BLEEDING IN SCALP ABOVE THE EAR
LIGHT PRESSURE IN FRONT OF THE MIDDLE EAR

BLEEDING ON OUTSIDE OR INSIDE OF HEAD
MODERATE PRESSURE ON NECK ABOUT 3" BELOW EAR AND 3" ABOVE COLLARBONE PUSH ARTERY AGAINST SPINE

BLEEDING IN THE CHEEK
VERY LIGHT PRESSURE IN NOTCH UNDER EDGE OF JAW 2/3 BACK FROM TIP OF CHIN

BLEEDING IN THE LOWER ARM
STRONG PRESSURE ON INSIDE OF ARM HALF WAY BETWEEN SHOULDER AND ELBOW

BLEEDING IN ARM
FIRM PRESSURE BEHIND MIDDLE OF COLLARBONE PUSH ARTERY AGAINST FIRST RIB

BLEEDING ABOVE KNEE
STRONG PRESSURE IN THE GROIN WITH HEEL OF HAND PUSH ARTERY AGAINST PELVIC BONE

Figure 51. Pressure Points for Bleeding

Elevate the limb if bleeding does not stop.

A tourniquet should be applied only for life-endangering hemorrhage that cannot be controlled by any other means, such as a pressure bandage. The tourniquet should be placed as close as possible to the wound on the other side toward the trunk.

No longer is it considered good first-aid practice to release applied tourniquets periodically at 15 to 30 minute intervals.

Once applied, a tourniquet should not be released regardless of the time interval elapsed, except by someone who is prepared to control the hemorrhage by other means, and to replace blood volume adequately. It is readily apparent, therefore, that the use of a tourniquet is indicated only in extremely serious cases, such as complete severance of a limb. Try whenever humanly possible to rely on the use of pressure to control bleeding.

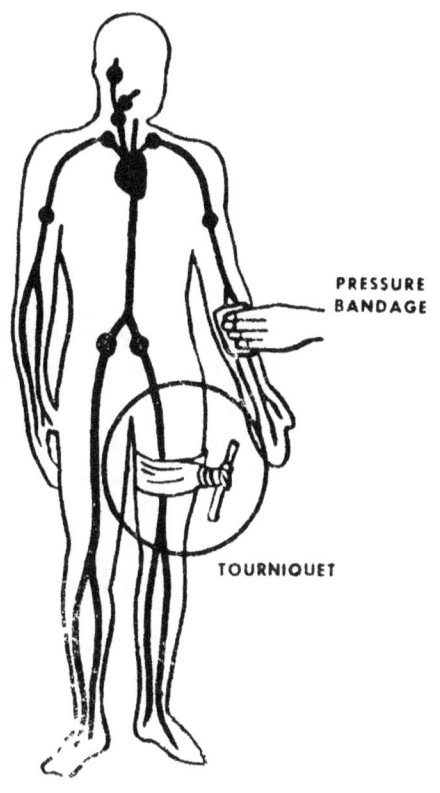

BLEEDING BELOW THE KNEE AND ELBOW
USE TOURNIQUET BETWEEN
CROTCH AND KNEE

Figure 52. Pressure Points for Bleeding

Cessation of Breathing

If breathing has stopped, apply artificial respiration at once. Be sure that the patient's tongue is pulled forward.

Shoulder-Blade (Back) Pressure Armlift Method of Artificial Respiration.

1. Place the patient face down with his head resting on hands.

2. Open his mouth, remove foreign articles, including false teeth, and make sure that tongue is forward.

3. Kneel at patient's head. Place index fingers on lower ribs. Rock forward until your arms are vertical.

4. Roll off to the outside with a snap and grasp patient's elbows. Rock back on heels, release elbows smartly.

5. Repeat above rhythmically 12 times each minute.

Keep up artificial respiration at a normal rate until breathing is restored or patient is unquestionably dead (listen for heart beat with ear against his bare chest). Keep him quiet when breathing starts. If you have oxygen, give it in alternate 5-minute periods after breathing starts. Do not give the oxygen with the tank valve wide open.

Head Injuries

Look for head injuries or fractured skull (indicated by

Figure 53. Artificial Respiration

unequal pupils of the eye, bleeding from ears or into skin around eyes). Keep patient warm and dry, and handle gently. <u>Don't give morphine to men with head injuries.</u>

Chest Wounds

If an injured person has an open chest wound, through which air can be heard sucking, cover the wound with a large dressing. Air entering the wound will collapse the lungs; consequently, the pad should be firmly applied at the moment of maximum exhalation, just before more air is sucked in. It should be firm enough to make a seal but not tight enough to stop chest movement entirely.

Shock

All personnel will suffer some shock after an emergency landing. Men in shock may have pale, cold skin; they may sweat, breathe rapidly, and have a weak pulse; they may be confused or unconscious.

Lay the patient down flat on his back. Raise the feet, unless contraindicated (head injuries, breathing difficulties, flushed face or bleeding from ears, nose or mouth).

Keep him warm but not overheated. If he is conscious and not injured internally, give him warm drinks; do not give alcohol.

If oxygen is available give it to the patient.

If the patient is in severe pain from injury, give a morphine injection (syrette) according to directions on the container. (Always give the injection above the tourniquet, or on the uninjured extremity.)

Be reassuring and cheerful with men in shock.

Eye Injuries

Clean the wound and the eye by washing with sterile water. Use atrophine ointment or antibiotic ointment such as penicillin, terramycin, etc., when available. Cover the eye with sterile dressing. Give aspirin for pain.

To remove a foreign body in the eye, first irrigate with sterile water. If not successful, then wind sterile cotton on a match stick to make an applicator. Moisten with sterile water and attempt to dislodge the foreign body by several gentle swipes over the affected area. If this is unsuccessful, make no further attempt to remove it but use atrophine ointment and antibiotics.

Urine is a primitive but effective source of sterile washing solution for eye injuries.

Fractures

Handle injured men with care to avoid causing them more injury. Splint them where they lie.

Don't attempt to remove clothing from a fractured limb in the normal manner. If a wound exists, cut or tear away clothing and treat before splinting. Clothing tears best at the seams.

Improvise splints from pieces of equipment or from a tight roll of clothing; pad with soft materials. The splint should be long enough to support the joints above and below the fracture.

Give a morphine injection (syrette) for severe pain (except for head injuries).

Keep the patient lying quiet; don't move him.

Sprains

Bandage and keep sprained part at rest. Cold localized applications may prevent swelling. When swelling has decreased (in 6 to 8 hours) localized application of heat will ease pain. Elevate the injured extremity.

If it is necessary to use the sprained limb, immobilize the injured area as much as possible with a splint or heavy wrapping. If only sprained, a limb can be used to the limit that pain will permit.

www.ingramcontent.com/pod-product-compliance
Lightning Source LLC
Chambersburg PA
CBHW031146160426
43193CB00008B/271